Claus Wilhelm Turtur

Conversion of the Zero-point Energy of the Quantum Vacuum into Classical Mechanical Energy

Claus Wilhelm Turtur

Conversion of the Zero-point Energy of the Quantum Vacuum into Classical Mechanical Energy

www.eh-verlag.de

Turtur, Claus Wilhelm
Conversion of the Zero-point Energy of the Quantum Vacuum into Classical Mechanical Energy

1. Auflage 2010
ISBN: 978-3-941482-60-9

www.eh-verlag.de

Bibliografische Informationen der Deutschen Nationalbibliothek:
Die Deutsche Nationalbibliothek verzeichnet diese Publikation in der Deutschen Nationalbibliografie; detaillierte bibliografische Daten sind im Internet über http://dnb.d-nb.de abrufbar.

EUROPÄISCHER
HOCH
SCHUL
VERLAG

Table of Contents

1. Introduction

The name "vacuum" is usually given to the space, out of which nothing can be taken with known methods. But it is well-known, that this vacuum is not empty, but it contains physical objects [Man 93], [Köp 97], [Lin 97], [Kuh 95]. This is also reflected within the Theory of General Relativity, namely by the cosmological constant Λ, which finally goes back to the gravitative action of the "mere space" [Goe 96], [Pau 00], [Sch 02]. Its name "cosmological constant" indicates, that the universe contains huge amounts of space, which lead to measureable effects, namely it influences the universe's rate of expansion [Giu 00], [Rie 98], [Teg 02], [Ton 03], [e1]. The crucial question of course is, whether it is possible to develop new methods, which allow to extract something from the vacuum, which could not be extracted up to now – some of those objects not visible directly up to now.

Already from the mass-energy-equivalence it is known, that the physical objects within the vacuum correspond with a certain amount of energy. This leads to the question, whether the "vacuum-energy" (i.e. the "energy of the empty space") can be made manifest in the laboratory. This question was answered positively in the work presented here. The description of the work begins with an explanation of the theoretical concepts in the sections 2 and 3, followed by an experimental verification in section 4, which describes the successful conversion of vacuum-energy into classical mechanical energy. Thus, the presented work introduces a new method to extract energy from the vacuum.

The described energy conversion raises the hope, that vacuum-energy can be used to supply mankind with energy, because it provides possibility to get energy from the immense amount of space which forms the universe, and which is large enough, that mankind will not be able to exhaust it. First of all, this source of energy is free from any pollution of the environment or from causing any damage to our habitat, the earth. Thus, in section 5 there are following some thoughts, regarding the future development of the energy-conversion method up to technical maturity.

2. Philosophical background

The crucial question to authenticate vacuum-energy was: "By which means is it possible to convert vacuum-energy into a classical form of energy (in order to make it visible) ?"

The way to the solution of this question is the following:

If vacuum-energy shall be converted into mechanical energy, there must be some mechanical forces. Responsible for the creation of such forces has to be one of the fundamental interactions of nature, as there are

- Gravitation,
- Electromagnetic interaction,
- Strong interaction,
- Weak interaction.

For Gravitation is not very strong on the one hand, and Strong and Weak interaction are rather difficult to operate, it was clear from the very beginning, that the most hopeful way for the conversion of vacuum-energy is via Electromagnetic interaction. This way was favoured especially after seeing the considerations published in [e1, e2, e3]. On this basis, the considerations following in section 2 have been developed.

2.1. Static fields versus Theory of Relativity

Within Classical Electrodynamics, there is no speed of propagation being attributed to electrostatic fields same as to magnetostatic fields (as far as DC-fields are under consideration, not AC-fields or waves) [Jac 81], [Gre 08]. But rather such fields are regarded as existing everywhere in the space at the same time, at each position with its appropriate field strength, which is calculated for electric fields by the means of Coulomb's law and for magnetic fields by the means of Biot-Savart's law – but without taking the speed of propagation into consideration as long as we follow conventional classical point of view.

This conventional point of view is in sharp contradiction with the Theory of Relativity, according to which the speed of light is a principal upper limit for all types of speed at all. If we accept the Theory of Relativity, we have to accept finite speed of propagation of electric and magnetic fields (also for DC-fields).

The contradiction between Classical Electrodynamics and the Theory of Relativity can be illustrated with the following example:

Please imagine the process of electron-positron pair-production, at which a photon decays into an electron and a positron. This process acts in separating electrical charges, which now (after they are created) produce electric fields because of their existence and magnetic fields because of their movements. Following the conception of Classical Electrodynamics, these fields should be observable everywhere in the space immediately after the moment at which the pair-production had happened, because the finite speed of light should only be applied to the propagation of electromagnetic waves but not to the propagation of DC-fields. If this approach would be appropriate, it would be possible to transport information with infinite speed (which is much faster than the speed of light), just by moving electrical charges and measuring the produced electrostatic field strength somewhere in the space [Chu 99], [Eng 05]. It is obvious, that this is in clear contradiction with the Theory of Relativity.

However there is a conception with Electromagnetic Field Theory, exceeding this simple view of Classical Electrodynamics, namely the retarded potentials of Liénard and Wiechert, which finally go back to the fundamental explanation, that the four-dimensional potentials of moving charge configurations follow a finite speed of propagation with their fields and field strengths. Each four-dimensional Liénard-Wiechert-potential can be subdivided into a three-dimensional vector potential plus a one-dimensional scalar potential [Kli 03], [Lan 97]. Based on this conception, the electric field strength corresponding with the one-dimensional scalar potential was calculated in the present work, for the example of a configuration of several moving point charges, and the result was plotted as a function of ongoing time at a given position. The purpose of this calculation is to demonstrate, that the time dependent field strength at a given position is indeed dependant on the questions, whether the finite speed of propagation of the (DC-) field is taken into account or not [e4]. We want to do this now:

We begin our consideration of the field strength with Coulomb's law [Jac 81], which tells us the force $\vec{F}$ between two point charges according to the expression

$$\vec{F} = \frac{q_1 \cdot q_2 \cdot \vec{r}}{4\pi\varepsilon_0 \cdot |\vec{r}|^3} \qquad (1.1)$$

with q_1, q_2 = electrical point charges,
$\vec{r}$ = distance between both charges,
$\varepsilon_0 = 8.854187817 \cdot 10^{-12} \frac{A \cdot s}{V \cdot m}$ = electrical field constant [Cod 00].

In this expression we have the option to choose which of the both electrical charges (for instance no.1) we want to interprete as the source of the field within which the other electrical charge (this would then be no.2) does experience the force $\vec{F} = \vec{E} \cdot q_2$. The field strength $\vec{E}$ of the point charge No.1 under this interpretation is

$$\vec{E} = \frac{q_1 \cdot \vec{r}}{4\pi\varepsilon_0 \cdot |\vec{r}|^3} \quad (1.2)$$

with $\vec{r}$ = distance relatively to charge No.1

This field strength can be interpreted in two different manners by principle, namely

(a) in the sense of Classical Electrodynamics, which does not take the speed of propagation of the field into consideration, but rather assumes an instant propagation of the fields (corresponding with an infinite speed of propagation), leading to the consequence that every alteration of the electrical charge q_1 or of its position will be noticed everywhere in the space at the same moment, or

(b) in the sense of Electromagnetic Field Theory and the Theory of Relativity, according to which also every field strength needs a certain amount of time t to pass the distance $|\vec{r}|$, being calculated as

$$t = \frac{|\vec{r}|}{c} \quad (1.3)$$

with c = speed of light.

This means that the field produced by every field source will act at the distance $|\vec{r}|$ from the source with a delay time of t.

Fact is: The conception of (a.) is only an approximation (for small delay time t) and the conception of (b.) is not really unusual. This can be seen from the fact, that for instance the mechanism of the Hertz'ian dipole emitter is often explained on the basis of the conception (b.), taking into account the electrostatic field as well as the magnetic field [Ber 71].

Another argument to accept this restriction of the propagation of the fields to the speed of light for all fields of fundamental interaction is given in connection with gravitational waves. Their genesis can be explained by moving bodies (gravitating masses) which emit fields of gravitation propagating with the speed of light. This is a topic in several measurements actually being in progress now [Abr 92], [Ace 02], [And 02], [Bar 99], [Wil 02].

This means, that every electrical charge permanently emits electrical field, and this field propagates through the space with finite speed after it is emitted. The propagation of the field is not influenced by the position of the charge any further as asoon as the field is away from its source. So every charge emits DC-field permanently even if the charge is moving during time, and as soon as this field is sent into the space, an alteration of the position of the charge does not alter this part of the field any more, which already left the charge. This means: If a field source (for instance a charge) is moving in the space during time, it always emits field from the actual position where it is in time and space, so that the field will be observed continuously coming from a changing position.

The difference between the two conceptions of (a.) and (b.) shall be demonstrated now with a simple example of calculation, based on a configuration of four point charges. This is already enough to show the difference between both conceptions. The paths (positions as a function of time) of the four point charges of our example shall be given in (1.4):

$$\begin{aligned}
s_1(t) &= -\frac{d_1+d_2}{2} + \frac{d_2-d_1}{2}\cdot\cos\left(2\pi\cdot\frac{t}{\tau}\right) && \text{for a negative charge } q_1^{(-)}, \\
s_2(t) &= -\frac{d_1+d_2}{2} - \frac{d_2-d_1}{2}\cdot\cos\left(2\pi\cdot\frac{t}{\tau}\right) && \text{for a positive charge } q_2^{(+)}, \\
s_3(t) &= +\frac{d_1+d_2}{2} + \frac{d_2-d_1}{2}\cdot\cos\left(2\pi\cdot\frac{t}{\tau}\right) && \text{for a positive charge } q_3^{(+)}, \\
s_4(t) &= +\frac{d_1+d_2}{2} - \frac{d_2-d_1}{2}\cdot\cos\left(2\pi\cdot\frac{t}{\tau}\right) && \text{for a negative charge } q_4^{(-)},
\end{aligned} \tag{1.4}$$

The absolute values of our four charges shall be identical, so that the charges only differ in the algebraic sign: $\left|q_1^{(-)}\right| = \left|q_2^{(+)}\right| = \left|q_3^{(+)}\right| = \left|q_4^{(-)}\right|$. The paths of the oscillations of those four charges are illustrated in fig.1.

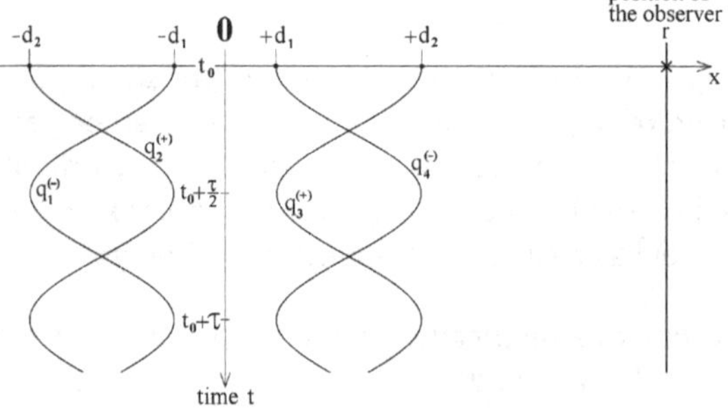

Fig. 1: Illustration of the oscillation of four charges during time as the given conditions for an example, which will be calculated in the following lines in order to demonstrate how the finite speed of propagation acts on time dependant development of the field strength.

Regarding the conception of (a.): The instantaneous propagation of the fields (with infinite speed) allows the superposition of the field strength simply linearly along the x-axis, following Coulomb's law. The consequence is, that the total charge configuration (of all four charges) does not produce any field at all (along the x-axis), so that a person observing the field (at the x-axis) will always come to the measurement of $\vec{E}=\vec{0}$.

The reason is:

The fact that the field strength is zero (along the x-axis) can be understood even without a calculation just by enhancing the dimensionality of the example to three dimensions. Than we would have two periodically contracting and expanding spherical shells (one with positive charge "+q" and the other one with negative charge "-q"), of which fig.2 shows a two-dimensional cut in the plane of the paper. And it is well known, that a charged sphere produces the same electrostatic field on its outside as a point charge in the middle of the sphere. But this statement is valid for both spheres (the positively charged sphere as well as the negatively charged sphere), so that both spheres produce a constant (DC-)field independently from the alteration of their radii during time. But in our example both spheres produce fields with the same absolute values but with the opposite algebraic signs, because both sphere have the same centre point. This means that the field strengths of both spheres compensate each other exactly to zero for all time. And of course, this consideration remains valid, if the dimensionality is reduced to one as shown in fig.1. Thus it is clear, that an oscillating charge configuration according to (1.4) and fig.1 does not produce any electrostatic field along the x-axis – as long as instant propagation of the field (with infinite speed) is assumed.

Remark: The thickness Δd of the spherical shells has been regarded as $\Delta d \rightarrow 0$ in our example.

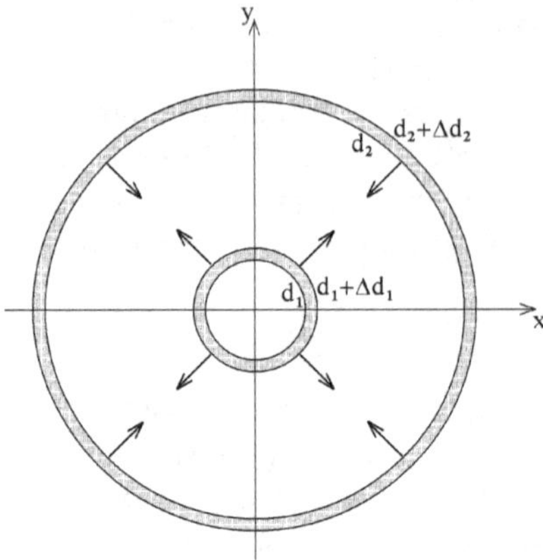

***Fig. 2:** Illustration of two electrically charged spherical shells, contracting and expanding periodically. In the moment displayed here, the outer shell (no.2, carrying the charge $-q$) contracts and the inner shell (no.1, carrying the charge $+q$) expands until they will pass through each other, and finally until d_2 will be smaller than d_1. The process of periodical contraction and expansion continues during all time of our observation.*
The image shows a 2-dimensional projection of a 3-dimensional assembly of spherical shells, representing their borders.

By the way, it shall be mentioned, that it is possible to calculate the field strength $\vec{E}$ at the position $\vec{x}$ in the one-dimensional case by the superposition of the fields of the four charges as written in (1.5), because the charges follow the conditions $q_1^{(-)} = q_4^{(-)} = -q$ and $q_2^{(+)} = q_3^{(+)} = +q$ (as presumed for our example), so that the vectors can be replaced by scalars in the one-dimensional case (with $\vec{s}$ = positions of the four charges):

$$\begin{aligned}\vec{E}_{ges} &= \vec{E}_1 + \vec{E}_2 + \vec{E}_3 + \vec{E}_4 \\ &= \frac{q_1 \cdot (\vec{x} - \vec{s}_1)}{4\pi\varepsilon_0 |\vec{x} - \vec{s}_1|^3} + \frac{q_2 \cdot (\vec{x} - \vec{s}_2)}{4\pi\varepsilon_0 |\vec{x} - \vec{s}_2|^3} + \frac{q_3 \cdot (\vec{x} - \vec{s}_3)}{4\pi\varepsilon_0 |\vec{x} - \vec{s}_3|^3} + \frac{q_4 \cdot (\vec{x} - \vec{s}_4)}{4\pi\varepsilon_0 |\vec{x} - \vec{s}_4|^3} \\ &\text{and for the one-dimensional case} \\ &= \frac{-q}{4\pi\varepsilon_0 \cdot (x - s_1)^2} + \frac{q}{4\pi\varepsilon_0 \cdot (x - s_2)^2} + \frac{q}{4\pi\varepsilon_0 \cdot (x - s_3)^2} - \frac{q}{4\pi\varepsilon_0 \cdot (x - s_4)^2} = \vec{0}\end{aligned} \tag{1.5}$$

Regarding the conception of (b.): If the finite speed of propagation of the fields is taken into account, the field strength is totally different. Already along the x-axis a field strength different from zero can be observed. The reason is that fields coming from different point charges (i.e. from different positions on the spherical shells) have to pass different amounts of distance and thus time to reach the observer. We see this in the following lines:

Of course the field strength of each point charge no. i (with $i=1...4$) at the position x is given by Coulomb's law (in analogy with $|\vec{E}|_1,...,|\vec{E}|_4$ at (1.5)), but the addition of all four field strengths requires the consideration of the time at which it was produced (i.e. at which it was emitted from the charge). Thus, the calculation has to take the moment of the emission of the field into account as well as the duration, which the fields need to propagate from the source to the observer, travelling with the speed of light c. One the basis of (1.3), this duration can be calculated as according to (1.6).

$$\Delta t_i = \frac{x - s_i(t)}{c} \qquad \text{(with } i=1...4\text{)}. \qquad (1.6)$$

If the calculation of the field strength is conducted with continuously ongoing time and the superposition of those field strengths which reach the observer at the position x at the same moment (they have to be calculated individually for each of the four point charges), the result will be the total electric (DC-)field at the position x as a function of time, following Coulomb's law under consideration of the finite speed of propagation of the field.

An exemplary result, calculated with the input-values of $q=1.60217653\cdot 10^{-19}\text{C}$ (elementary charge), $d_1=0.5m$, $d_2=3.5m$, $\tau=10^{-7}\text{sec.}$ and $x=10m$, is plotted in fig.3. Obviously, the field strength at the position of the observer is not zero. This is plausible, because those field strengths which compensate each other in (1.5) will not compensate to zero now, because they arrive at the position x at different moments in time, because they have to pass different distances on their way to the observer. On the other hand those field-strengths which reach at the position x have been produced at different moments and thus at different positions not being able to compensate each other according to (1.5), because there are different distances to be used in Coulomb's law.

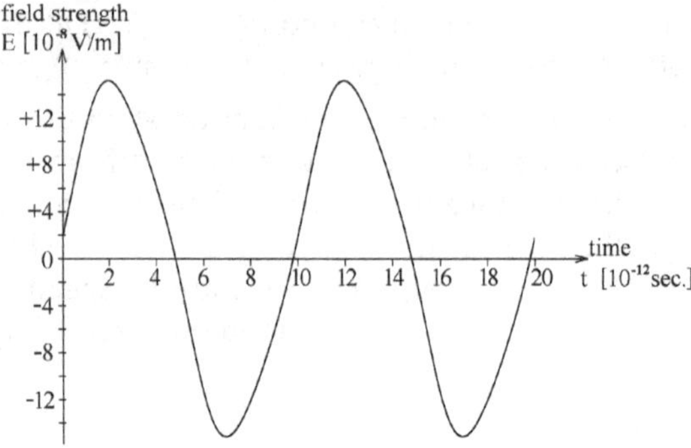

Fig. 3: Examplary result of the electrostatic field strength generated by four charges oscillating as given in fig.1 and in (1.4). The calculations are based on Coulomb's law with taking additionally the finite speed of propagation of the fields into account.

The most important statement of the result is obvious: The field strength is not zero at the position x.

But there is a further observation: Although the four point charges oscillate following a cosine during time, the field strength (at the position x) as a function of time does not exactly follow a sine or cosine. Its deviation from a sine can be seen in equation (1.7), where the first sine-term gives the dominant part of the field strength, and the next both sine-terms give the continuation of a series (which is not a Fourier-series because of the individual phase of every sine-term). By the way: The maximum of the speed, with which the electrical charges move in our example is a bit less than $2 \cdot 10^{8} \, m/s$. If the charges move slower, the difference between the shape of $E(t)$ and a sine decreases.

The field strength is (calculated with an accuracy of $\Delta E = \pm 3 \cdot 10^{-14} \frac{V}{m}$)

$$E(t) = a_0 \cdot \sin\left(1 \cdot \frac{2\pi}{\tau} \cdot (t - a_1)\right) + a_2 \cdot \sin\left(3 \cdot \frac{2\pi}{\tau} \cdot (t - a_3)\right) + a_4 \cdot \sin\left(5 \cdot \frac{2\pi}{\tau} \cdot (t - a_5)\right) + \ldots \tag{1.7}$$

with the coefficients $a_0 = -1.47671114257737 \cdot 10^{-11} \frac{V}{m}$, $a_1 = 4.68214978523477 \cdot 10^{-8} \text{ sec.}$,

$a_2 = -9.46983556843000 \cdot 10^{-13} \frac{V}{m}$, $a_3 = 5.65583264190749 \cdot 10^{-8} \text{ sec.}$,

$a_4 = +6.47000984663877 \cdot 10^{-14} \frac{V}{m}$, $a_5 = 4.67175185668795 \cdot 10^{-8} \text{ sec.}$

The consequence for the present work is the following:

As soon as we take the conception of (b.) following the Electromagnetic Field Theory as well as the Theory of Relativity serious, we accept, that it is possible to construct a charge configuration, which produced an electrical field only because of the finite speed of propagation of the field strength. This gives us a criterion to decide (by experiment) whether the classical conception (a.) is correct or the modern conception (b.), namely as following: It should be possible to find a charge configuration, which produced electrostatic forces – but only if the conception (b.) is correct, but not producing any fields and forces according to the conception (a.).

Of course, such forces do not have an explanation within Classical Electrodynamics – they do not exist within this Theory. Consequently, they should have an explanation within Electromagnetic Field Theory or within Quantum Electrodynamics (going back to the structure of the vacuum / space). From this point of view it can be concluded, that the finite speed of propagation of electric (DC-)fields (as well as of magnetic (DC-)fields) should offer a possibility to convert some of the energy within the vacuum (space) into some classical type of energy – as will be demonstrated in the further course of the present work. (The logics behind this conclusion will be seen soon.)

Additional remark:

The contradictions between Classical Electrodynamics on the one hand and Electromagnetic Field Theory and the Theory of Relativity on the other hand can be solved in favour of Field Theory and Relativity – if it is possible to conduct an experiment, which converts vacuum-energy into classical energy, which is developed and explained on the basis of the finite speed of propagation of the field. This will be done in the experimental part of the present work. From this point of view it is clear, that Classical Electrodynamics is an approximation under terrestric circumstances for normal technical applications, where the finite speed of propagation of electric and magnetic fields will not be recognized, because the distances inside a laboratory, which the fields have to transit are so small, that the duration of propagation is not noticed. This is not astonishing if we have a look to the time scale in fig.3 (Pikoseconds) in comparison distances of some meters ($x = 10\,Meter$).

But there is one additional question arising: From where does the energy originate, which the propagating field contains and transports ?

This question is important and helpful, because the mentioned energy is this type of energy, which shall be converted into a classical type of energy. We want to trace this energy exemplarily in section 2.2 for the electric field and in section 2.3 for the magnetic field. This leads us to the logical background, from which was said, that it will understood soon.

2.2. A circulation of energy of the electrostatic field

If electrostatic fields propagate with the speed of light, they transport energy, because they have a certain energy density [Chu 99]. It should be possible to trace this transport of energy if is really existing. That this is really the case can be seen even with a simple example regarding a point charge, as will be done on the following pages. When we trace this energy, we come to situation, which looks paradox at the very first glance, but the paradox can be dissolved, introducing a circulation of energy [e5]. This is also demonstrated on the following pages. By the way there are colleagues who also began with considerations of this propagating energy (for instance [Eng 05]), but they do not primarily discuss the circulation of energy but rather the idea to transmit information with a speed faster than the speed of light, which could only be possible according to the conception of the infinite speed of propagation of electric (or magnetic) DC-fields.

The first aspect of the mentioned paradox regards the emission of energy at all[1] [e16]. If a point charge (for instance an elementary charge) exists since a given moment in time, it emits electric field and field's energy from the time of its birth without any alteration of its mass. The volume of the space filled with this field increases permanently during time and with it the total energy of the field. But from where does this "new energy" originate ? For the charged particle does not alter its mass (and thus its energy), the "new energy" can not originate from the particle itself. This means: The charged particle has to be permanently supplied with energy from somewhere. The situation is also possible for particles, which are in contact with nothing else but only with the vacuum. The consequence is obvious: The particle can be supplied with energy only from the vacuum. This sounds paradox, so it can be regarded as the first

---

[1] For the sake of illustration it should be mentioned that periodically moving stars (for instance such as rotating double stars [Sha 83] or a star rotating around a black hole) emit gravitational waves because of the emission of gravitational (DC-)fields. The fact that we see gravitational waves (at our position) goes back to the mechanism described in section 2.1 for moving point charges in similar way as for moving point masses. In this context it might be remembered, that this is not the gravimagnetic field known from the Thirring-Lense-effect (see for instance [Sch 02], [Thi 18], [Gpb 07]), which can be understood in analogy with the magnetic field of Elecrodynamic and which is emitted additionally.

aspect of the mentioned paradox. But it is logically consequent, and so we will have to solve it later.

Remark: Electrically charged particles can be "born" by pair production process, but there might also be charged particles existing since the big bang. Both types of particles are well within the explanations given here. The difference is only the duration of their existence, which is proportional to the diameter of the sphere, which is filled with the field. But for our paradox, this difference does not play any role.

The second aspect of the mentioned paradox regards the propagation of the emitted field's energy into the space. In order to understand this, we want to regard the electrical field emitted from a charged elementary particle Q (see fig.4). For the our consideration it does not play a role, whether the elementary particle has punctiform shape (for instance as the electron in scattering experiments, leading to a radius of $r_{streu} < 10^{-18} m$, see for instance [Loh 05], [Sim 80]) or whether we regard an electron with its classical radius ($r_{klass.} = 2.82... \cdot 10^{-15} m$ according to [Cod 00], [Fey 01]). In order to evade such questions, we want to put a sphere with the radius x_1 around the particle Q. Furthermore we want to fix our time scale at this moment $t = 0$, at which the electrostatic field fills exactly the sphere with the radius x_1. From there on, we trace the field along its propagation through the space. Let us now come to a moment $\Delta t > 0$, which is later than the begin of our consideration, so that the field with its finite speed of propagation c fills a sphere with the radius $x_1 + c \cdot \Delta t$, so that the energy which the charge emitted during the time-interval Δt is the energy within the spherical shell from x_1 to $x_1 + c \cdot \Delta t$, because this is the amount of energy, by which the total energy of the field was enhanced during the time interval Δt. This energy is larger than zero, so that we clearly see that the charge Q indeed emitted field and field's energy.

Let us now observe the situation at a moment t_2, which is later than the moment in the consideration before. And let us further add the time interval Δt, so that the spherical shell from x_1 to $x_1 + c \cdot \Delta t$ had developed itself into the spherical shell from x_2 to $x_2 + c \cdot \Delta t$. This means that the energy from the shell $x_1 \; ... \; x_1 + c \cdot \Delta t$ moved into the shell $x_2 \; ... \; x_2 + c \cdot \Delta t$. If the vacuum (inside which the field propagates) would not extract energy from the field, we expect the energy within the shell $x_1 \; ... \; x_1 + c \cdot \Delta t$ to be the same as the energy within the shell $x_2 \; ... \; x_2 + c \cdot \Delta t$.

Will now check this expectation by the following calculations and we will find a violation of the energy conservation during the propagation of the field into the space. We will see that the field loses energy during its propagation. With other words: We regard a given package of space filled with electric field (inside a spherical shell) and trace it along its propagation through the space. We calculate the field's energy inside the field package and we will find that it does not keep its field's energy constant. This is the second aspect of the mentioned paradox of the electric field, which will also be solved later in the present work.

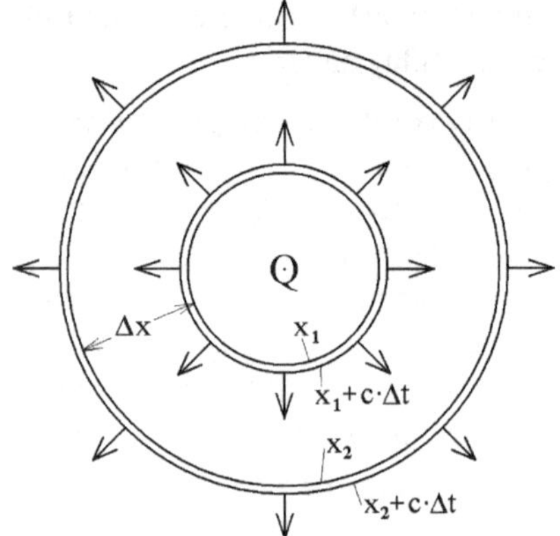

Fig. 4: Illustration of a spherical shell, which contains a certain amount of field energy of an electrostatic field. The sense of this construction is to trace the field energy when passing the empty space.

The trace of the propagating energy within the spherical shell during time and space is calculated as following:

- The field strength produced by a charge Q with radial symmetry (i.e. a punctiform charge or a charge with spherical symmetry) is according to Coulomb's law

$$\vec{E}(\vec{r})=\frac{1}{4\pi\varepsilon_0}\cdot\frac{Q}{r^3}\cdot\vec{r}\,, \tag{1.8}$$

  where the centre of the charge is located in the origin of coordinates and $\vec{r}$ is the position vector of an arbitrary point in the space at which the field strength shall be determined.

- If we write $\vec{r}$ in spherical coordinates with $\vec{r}=(r,\vartheta,\varphi)$, the absolute values of the field strength are dependant not of the direction of $\vec{r}$ but only of the absolute value of $r=|\vec{r}|$, namely

$$E=\left|\vec{E}\right|=\frac{1}{4\pi\varepsilon_0}\cdot\frac{Q}{r^2}\,. \tag{1.9}$$

- The energy density of the electric field is

$$u=\frac{\varepsilon_0}{2}\cdot\left|\vec{E}\right|^2. \tag{1.10}$$

- Consequently the energy density of the field produced by a charge with spherical symmetry is

$$u=\frac{\varepsilon_0}{2}\cdot\left|\vec{E}\right|^2=\frac{\varepsilon_0}{2}\cdot\left(\frac{1}{4\pi\varepsilon_0}\cdot\frac{Q}{r^2}\right)^2=\frac{Q^2}{32\pi^2\varepsilon_0 r^4} \tag{1.11}$$

- The energy within the spherical shell from x_1 to $x_1+c\cdot\Delta t$ can now be calculated as the Volume integral

$$\begin{aligned}
E_{\substack{inner\\shell}} &= \int\limits_{\substack{spherical\\shell}} u(\vec{r})\,dV = \int\limits_{\varphi=0}^{2\pi}\int\limits_{\vartheta=0}^{\pi}\int\limits_{r=x_1}^{x_1+c\cdot\Delta t}\frac{Q^2}{32\pi^2\varepsilon_0 r^4}\cdot r^2\cdot\sin(\vartheta)\,dr\,d\vartheta\,d\varphi \\
&= \frac{Q^2}{32\pi^2\varepsilon_0}\cdot\int\limits_{\varphi=0}^{2\pi}\int\limits_{\vartheta=0}^{\pi}\underbrace{\int\limits_{r=x_1}^{x_1+c\cdot\Delta t}\frac{1}{r^2}\cdot dr}_{=\frac{c\cdot\Delta t}{(x_1+c\cdot\Delta t)\cdot x_1}}\cdot\sin(\vartheta)\,d\vartheta\,d\varphi \\
&= \frac{Q^2}{32\pi^2\varepsilon_0}\cdot\frac{c\cdot\Delta t}{(x_1+c\cdot\Delta t)\cdot x_1}\cdot\underbrace{\int\limits_{\varphi=0}^{2\pi}\underbrace{\int\limits_{\vartheta=0}^{\pi}\sin(\vartheta)\,d\vartheta}_{=2}\,d\varphi}_{=4\pi} \\
&= \frac{Q^2}{32\pi^2\varepsilon_0}\cdot\frac{c\cdot\Delta t}{(x_1+c\cdot\Delta t)\cdot x_1}\cdot 4\pi = \frac{Q^2}{8\pi\varepsilon_0}\cdot\frac{c\cdot\Delta t}{(x_1+c\cdot\Delta t)\cdot x_1}
\end{aligned} \tag{1.12}$$

Obviously, this energy is not zero. This means that the charge (which is the field source) indeed emits energy permanently. By the way, this is a mathematical reproduction of the first paradox.

- Let now the time elapse until it reaches t_2. The inner border of the observed spherical shell has now passed from x_1 to x_2 and the outer border from $x_1+c\cdot\Delta t$ to $x_2+c\cdot\Delta t$. With the distance Δx introduced in fig.4, we find the inner and the outer border of the shell being at the radii $x_2=x_1+\Delta x$ respectively $x_2+c\cdot\Delta t=x_1+\Delta x+c\cdot\Delta t$. The spherical shell has enhanced its volume, but the field strength within this moving shell has been reduced (in accordance with Coulomb's law). If the empty space would allow the field energy just to pass by, the amount of energy within the outer shell $E_{outer\ shell}$ should be the same as the amount of energy within the inner shell $E_{inner\ shell}$. We want to check this and we will find that this is not the case:

$$E_{\substack{outer\\shell}} = \int\limits_{\substack{spherical\\shell}} u(\vec{r})\,dV = \int\limits_{\varphi=0}^{2\pi}\int\limits_{\vartheta=0}^{\pi}\int\limits_{r=x_2}^{x_2+c\cdot\Delta t} \frac{Q^2}{32\pi^2\varepsilon_0 r^4}\cdot r^2\cdot\sin(\vartheta)\,dr\,d\vartheta\,d\varphi$$

$$= \frac{Q^2}{32\pi^2\varepsilon_0}\cdot\int\limits_{\varphi=0}^{2\pi}\int\limits_{\vartheta=0}^{\pi}\underbrace{\int\limits_{r=x_1+\Delta x}^{x_1+\Delta x+c\cdot\Delta t}\frac{1}{r^2}\cdot dr}_{=\frac{c\cdot\Delta t}{(x_1+\Delta x+c\cdot\Delta t)\cdot(x_1+\Delta x)}}\cdot\sin(\vartheta)\,d\vartheta\,d\varphi$$

$$= \frac{Q^2}{32\pi^2\varepsilon_0}\cdot\frac{c\cdot\Delta t}{(x_1+\Delta x+c\cdot\Delta t)\cdot(x_1+\Delta x)}\cdot\underbrace{\int\limits_{\varphi=0}^{2\pi}\underbrace{\int\limits_{\vartheta=0}^{\pi}\sin(\vartheta)\,d\vartheta}_{=2}\,d\varphi}_{=4\pi}$$

$$= \frac{Q^2}{32\pi^2\varepsilon_0}\cdot\frac{c\cdot\Delta t}{(x_1+\Delta x+c\cdot\Delta t)\cdot(x_1+\Delta x)}\cdot 4\pi = \frac{Q^2}{8\pi\varepsilon_0}\cdot\frac{c\cdot\Delta t}{(x_1+\Delta x+c\cdot\Delta t)\cdot(x_1+\Delta x)} \qquad (1.13)$$

- Obviously, the energy $E_{inner\ shell}$ is more than the energy $E_{outer\ shell}$. This means that the empty space decreases the energy of the shell. This proofs the validity of the second paradox, and we see that the vacuum (the mere space) takes away energy from the field during its propagation.

The energy loss of the field during its propagation can be calculated by subtraction of the energy contained in the shells according to (1.12) and (1.13) for the example of a point charge. The spherical shell propagating from $x_1 \ldots x_1+c\cdot\Delta t$ to $x_2 \ldots x_2+c\cdot\Delta t$ loses the energy

$$\Delta E = E_{\substack{inner\\shell}} - E_{\substack{outer\\shell}} = \frac{Q^2}{8\pi\varepsilon_0}\cdot\frac{c\cdot\Delta t}{(x_1+c\cdot\Delta t)\cdot x_1} - \frac{Q^2}{8\pi\varepsilon_0}\cdot\frac{c\cdot\Delta t}{(x_1+\Delta x+c\cdot\Delta t)\cdot(x_1+\Delta x)} \qquad (1.14)$$

For small but finite Δt and Δx (which we can neglect as summands in comparison with the non-infinitesimal x_1 in a mathematical limes, but which we can not neglect as factors), we come to the good approximation

$$\Delta E = E_{\substack{inner\\shell}} - E_{\substack{outer\\shell}} = \frac{Q^2}{8\pi\varepsilon_0}\cdot\frac{(2x_1+\Delta x+c\cdot\Delta t)\cdot c\cdot\Delta t\cdot\Delta x}{(x_1+c\cdot\Delta t)\cdot x_1\cdot(x_1+\Delta x+c\cdot\Delta t)\cdot(x_1+\Delta x)}$$
$$\approx \frac{Q^2}{8\pi\varepsilon_0}\cdot\frac{2x_1\cdot c\cdot\Delta t\cdot\Delta x}{x_1\cdot x_1\cdot x_1\cdot x_1} = \frac{Q^2}{8\pi\varepsilon_0}\cdot\frac{2\cdot c\cdot\Delta t\cdot\Delta x}{x_1^3}. \qquad (1.15)$$

As we see, the loss of energy decreases with the third potential of the radius of the shell x_1 (for a finite thickness of the shell $c\cdot\Delta t$, which can not be infinite, so that there will still be energy inside the shell) (and for a finite distance of propagation Δx, which can not be infinite, so that there is propagation at all). This leads us to the following systematic with regard to the distance r from the centre of the charge:

Field-parameter		Proportionality
Electric Potential of a point charge	→	$V \propto r^{-1}$
Field strength caused by a point charge (Coulomb's law)	→	$F \propto r^{-2}$
Dispersed energy of a spherical shell around a point charge	→	$\Delta E \propto r^{-3}$
Energy density of the field of a point charge	→	$u \propto r^{-4}$

Important is the conclusion, which can be found with logical consequence:

On the one hand the vacuum (= the space) permanently supplies the charge with energy (first paradox aspect), which the charge (as the field source) converts into field energy and emits it in the shape of a field. On the other hand the vacuum (= the space) permanently takes energy away from the propagating field, this means, that space gets back its energy from field during the propagation of the field. This indicates that there should be some energy inside the "empty" space, which we now can understand as a part of the vacuum-energy. In section 3, we will understand this energy more detailed.

But even now, we can come to the statement:

During time, the field of every electric charge (field source) increases. Nevertheless the space (in the present work the expressions "space" and "vacuum" are use as synonyms) causes a permanent circulation of energy, supplying charges with energy and taking back this energy during the propagation of the fields. This is the circulation of energy, which gave the title for present section 2.2.

This leads us to a new aspect of vacuum-energy:

The circulating energy (of the electric field) is at least a part of the vacuum-energy. We found its existence and its conversion as well as its flow. On the basis of this understanding it should be possible to extract at least a part of this circulating energy from the vacuum – in section 4 a description is given of a possible method how to extract such energy from the vacuum.

2.3. A circulation of energy of the magnetostatic field

Because of some similarities between the electrostatic field and the magnetostatic field (the last-mentioned can be led back to the first-mentioned by a Lorentz-transformation, see [Sch 88], [Dob 06]), it should be possible to find a circulation of field energy also for the magnetic field in analogy with the circulation of field energy as found for the electric field. In section 2.3 it is demonstrated, that this analogy is going rather far. We want to demonstrate this with an example, for which we have to chose the geometry of the field source a bit different from the example of the energy circulation of the electric field. The reason can be understood in connection with the Lorentz-transformation mentioned above, which has the consequence (among others) that a magnetic field can not have a punctiform field source, because the creation of a magnetic fields need the movement of electrical charge. (The lowest order of the magnetic multiple is the dipole and not the monopole as in the case of the electric field.) Thus, our example in section 2.3 shall be built up on an electrical charge moving with constant velocity, emitting a constant magnetic field. And we want to follow the propagation of this field into the space, similar as we did in section 2.2.

Let us start our example with a configuration of moving electrical charge, producing a magnetic field as drawn in fig.5. The charge shall be geometrically arranged homogeneously in a line with infinite length, orientated along of the z-axis, and the whole line is moving continuously in z-direction with constant speed. The absolute value of the magnetic field strength $H=|\vec{H}|$ can than be found in a usual standard textbook for students, for instance as [Gia 06]. It is

$$H=|\vec{H}|=\frac{I}{2\pi r} \quad \text{with } I=\text{electrical current and } r=\sqrt{x^2+y^2} \qquad (1.16)$$

The absolute value of the field strength is sufficient for the calculation of the energy density of the magnetic field, which also can be found in textbooks. It is

$$u = \frac{\mu_0}{2} \cdot \left| \vec{H} \right|^2 = \frac{\mu_0 I^2}{8\pi^2 r^2} \qquad \text{with } \mu_0 = 4\pi \cdot 10^{-7} \frac{V \cdot s}{A \cdot m} = \text{magnetic flux constant} \tag{1.17}$$

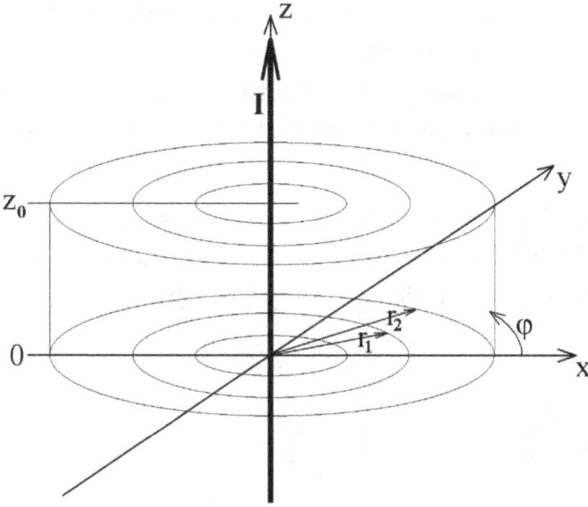

Fig. 5: Illustration of the homogeneous charge configuration along the z-axis with movement also along the z-axis. It produces the same magnetic field as an electrical conductor of infinite length. Because of the orientation of the current along the z-axis, the absolute value of the magnetic field strength can easily be given in cylindrical coordinates according to (1.16) and (1.17).

We now analyze the propagation of the magnetic field respectively of its energy in the space. This field together with its energy starts at the position of the z-axis and propagates perpendicularly to the z-axis (radially as shown in fig.5) with the speed of light. A component of propagation into the z-direction is not to be taken into account, because the whole setup is arranged with cylindrical symmetry around the z-axis (with infinite length in z-direction). This can also be understood, if we look to a volume element with the shape of a cylinder of finite length $z = 0 \ldots z_0$ (see fig.5). The energy flux through the top and through the bottom end, going into the cylinder and coming out of the cylinder are identically the same, because the cylinder is neither source nor sink. In this way, we understand, that the energy being emitted from the moving charge (which is located along the z-axis), is flowing with cylindrical symmetry into the xy-plane.

We now want to find out, how much magnetic field energy is flowing into the space within a time interval Δt. Therefore, we adjust the time-scale as following: The electrical current (i.e. the movement of the electrical charge) is switched on in the moment $t_0 = 0$. The time t_1 (with $t_1 > 0$) shall be defined as the moment, at which the magnetic field reaches the radius r_1 in consideration of its finite speed of propagation (see again fig.5). Again a bit later, namely at the moment $t_2 = t_1 + \Delta t$ (with $\Delta t > 0$), the field will reach a cylinder with the radius r_2. Consequently, the magnetic energy, which has been emitted by the moving charge within the time interval Δt, has to be the same energy, which fills the cylindrical shell from the inner radius r_1 up to the outer radius r_2.

We calculate this amount of energy by integration of the energy density inside the cylindrical shell (with finite height z_0), following equation (1.17) in which we introduce the magnetic field according to (1.16):

$$W = \iiint\limits_{\text{(cylinder)}} u\, dV = \int\limits_{\varphi=0}^{2\pi} \int\limits_{r=r_1}^{r_2} \int\limits_{z=0}^{z_0} \frac{\mu_0}{2} \cdot \left|\vec{H}\right|^2 \cdot r\, dz\, dr\, d\varphi = \int\limits_{\varphi=0}^{2\pi} \int\limits_{r=r_1}^{r_2} \int\limits_{z=0}^{z_0} \frac{\mu_0}{2} \cdot \frac{I^2}{(2\pi r)^2} \cdot r\, dz\, dr\, d\varphi$$

$$= \int\limits_{\varphi=0}^{2\pi} \int\limits_{r=r_1}^{r_2} \int\limits_{z=0}^{z_0} \frac{\mu_0 \cdot I^2}{8\pi^2 r} \cdot dz\, dr\, d\varphi = \int\limits_{\varphi=0}^{2\pi} \int\limits_{r=r_1}^{r_2} \frac{\mu_0 \cdot I^2}{8\pi^2 r} \cdot [z]_0^{z_0}\, dr\, d\varphi = \int\limits_{\varphi=0}^{2\pi} \int\limits_{r=r_1}^{r_2} \frac{\mu_0 \cdot I^2 z_0}{8\pi^2 r}\, dr\, d\varphi$$

$$= \int\limits_{\varphi=0}^{2\pi} \frac{\mu_0 \cdot I^2 z_0}{8\pi^2} \cdot \left(\ln(r_2) - \ln(r_1)\right) d\varphi = \frac{\mu_0 \cdot I^2 z_0}{8\pi^2} \cdot \left(\ln(r_2) - \ln(r_1)\right) \cdot 2\pi = \frac{\mu_0 \cdot I^2 z_0}{4\pi} \cdot \ln\left(\frac{r_2}{r_1}\right) \tag{1.18}$$

Furthermore, the emitted power can be calculated rather easy by dividing the emitted energy W through the time interval Δt within which this energy has been emitted. This time interval can be calculated from the finite speed of propagation (i.e. the speed of light) and the distance over which the magnetic field has propagated, namely as following:

$$c = \frac{r_2 - r_1}{\Delta t} \quad \Rightarrow \quad \Delta t = \frac{r_2 - r_1}{c} \tag{1.19}$$

Thus, we know, at which moments the radii r_1 and r_2 had been reached by the field:

$$c = \frac{r_1}{t_1} \quad \Rightarrow \quad r_1 = c \cdot t_1$$
$$\text{and} \qquad r_2 = c \cdot t_2 = c \cdot (t_1 + \Delta t) \tag{1.20}$$

This leads us to the information about the emitted power:

$$P = \frac{W}{\Delta t} = \frac{\mu_0 \cdot I^2 z_0}{4\pi \cdot \Delta t} \cdot \ln\left(\frac{r_2}{r_1}\right) \tag{1.21}$$

In order to find out, whether the power P is constant in time (as it should be expected in the case that the vacuum would not interact with the propagating field), we have to express the radii r_1 and r_2 as a function of time. For this purpose, we can use (1.20) and (1.21) and deduce:

$$P=\frac{W}{\Delta t}=\frac{\mu_0 \cdot I^2 z_0}{4\pi \cdot \Delta t}\cdot \ln\left(\frac{c\cdot(t_1+\Delta t)}{c\cdot t_1}\right)=\frac{\mu_0 \cdot I^2 z_0}{4\pi}\cdot\frac{1}{\Delta t}\cdot \ln\left(1+\frac{\Delta t}{t_1}\right) \qquad (1.22)$$

Obviously, this expression is not constant in time. If it would be constant in time, it would not depend on the time t_1. This means that the moving charge produces a magnetic field and thus it emits power, although the movement keeps constant speed. We understand this as the first aspect of the energy circulation (i.e. the first aspect of the solved paradox), regarding the emission of energy from a field source. But the moving (field producing) charge is in connection only with the vacuum, thus we have to find out the origin of the emitted energy (which will lead us soon to the second aspect of the energy circulation).

By the way, it shall be mentioned, that permanent magnets permanently emit field energy, which gives a very clear and simple explanation for the analogy with first aspect of the energy circulation of the electric field as shown in section 2.2.

But there is an additional further observation: Obviously, the emitted power is not constant in time, although there is no alteration of the field strength as a function of time. This explains the pendant for first aspect of the energy circulation (in analogy with the electric field), namely the fact that the vacuum takes energy out of the propagating field. This can be demonstrated by tracing a cylindrical volume (and the energy within this volume) along its propagation through the space, analyzing whether the energy inside this cylindrical volume remains constant during the propagation.

Therefore we follow the observed cylindrical shell with the inner radius r_1 and the outer radius r_2 for a further time interval $\Delta t_x > 0$. Within this time, the inner radius will be enlarged until it reaches $r_3 = r_1 + c\cdot\Delta t_x$ and the outer radius will be enlarged until it reaches $r_4 = r_2 + c\cdot\Delta t_x$. So we see the following development of the situation during time:

▪ At the moment $t_2 = t_1 + \Delta t$ our cylindrical shell had had the inner radius r_1 and the outer radius r_2, this means that we look to the same cylindrical shell as in the calculation of (1.18).

▪ At the moment $t_2 + \Delta t_x = t_1 + \Delta t + \Delta t_x$ this cylindrical shell from $r_1 \dots r_2$ has propagated radially into the space until it reaches the inner radius of r_3 and the outer radius of r_4.

The energy of this last mentioned cylinder can be calculated in both moments in analogy to equation (1.18):

For those both moments of observation we compare the energy inside the cylindrical shell according to (1.18) and we thus we come to the values of (1.23) and (1.24):

▪ At $t_2 = t_1 + \Delta t$ the shell contains the energy $W_{12} = \frac{\mu_0 \cdot I^2 z_0}{4\pi} \cdot \ln\left(\frac{r_2}{r_1}\right)$. (1.23)

▪ At $t_2 + \Delta t_x = t_1 + \Delta t + \Delta t_x$ the shell contains the energy

$$W_{34} = \frac{\mu_0 \cdot I^2 z_0}{4\pi} \cdot \ln\left(\frac{r_4}{r_3}\right) = \frac{\mu_0 \cdot I^2 z_0}{4\pi} \cdot \ln\left(\frac{r_2 + c \cdot \Delta t_x}{r_1 + c \cdot \Delta t_x}\right). \tag{1.24}$$

Obviously, both expressions are different. If we want to understand the time dependency of the energy, we put r_1 and r_2 from (1.20) into (1.23) and (1.24) and we come to:

$$\text{At } t_2 = t_1 + \Delta t \;\Rightarrow\; W_{12} = \frac{\mu_0 \cdot I^2 z_0}{4\pi} \cdot \ln\left(\frac{r_2}{r_1}\right) = \frac{\mu_0 \cdot I^2 z_0}{4\pi} \cdot \ln\left(\frac{c \cdot (t_1 + \Delta t)}{c \cdot t_1}\right)$$
$$= \frac{\mu_0 \cdot I^2 z_0}{4\pi} \cdot \ln\left(\frac{(t_1 + \Delta t)}{t_1}\right) = \frac{\mu_0 \cdot I^2 z_0}{4\pi} \cdot \ln\left(1 + \frac{\Delta t}{t_1}\right) \tag{1.25}$$

$$\text{At } t_2 + \Delta t_x = t_1 + \Delta t_x + \Delta t \;\Rightarrow\; W_{34} = \frac{\mu_0 \cdot I^2 z_0}{4\pi} \cdot \ln\left(\frac{r_4}{r_3}\right) = \frac{\mu_0 \cdot I^2 z_0}{4\pi} \cdot \ln\left(\frac{r_2 + c \cdot \Delta t_x}{r_1 + c \cdot \Delta t_x}\right)$$
$$\Rightarrow\; W_{34} = \frac{\mu_0 \cdot I^2 z_0}{4\pi} \cdot \ln\left(\frac{c \cdot (t_1 + \Delta t) + c \cdot \Delta t_x}{c \cdot t_1 + c \cdot \Delta t_x}\right) \tag{1.26}$$
$$= \frac{\mu_0 \cdot I^2 z_0}{4\pi} \cdot \ln\left(\frac{t_1 + \Delta t + \Delta t_x}{t_1 + \Delta t_x}\right) = \frac{\mu_0 \cdot I^2 z_0}{4\pi} \cdot \ln\left(1 + \frac{\Delta t}{t_1 + \Delta t_x}\right)$$

As expected, it is clear that (1.25) and (1.26) are different. And because we have traced the energy within an expanding cylindrical shell, we can conclude from that fact that $W_{12} \neq W_{34}$, that there is an exchange of energy between the field and the vacuum.

The question is now, whether this exchange of energy means, that the vacuum takes energy from the field or whether the vacuum gets energy from the field. This question is answered by a comparison of W_{12} and W_{34}: Because of $\Delta t_x > 0$ it must be $1+\frac{\Delta t}{t_1} > 1+\frac{\Delta t}{t_1+t_x}$ and thus we see the relation

$$\ln\left(1+\frac{\Delta t}{t_1}\right) > \ln\left(1+\frac{\Delta t}{t_1+\Delta t_x}\right), \qquad (1.27)$$

which means in consideration of (1.25) and (1.26) that $W_{12} > W_{34}$. Consequently, the magnetostatic field gives energy to the vacuum during its propagation.

This makes the analogy between the circulation of energy of the electric field and the magnetic field complete. In both cases the field source is supplied with energy from the vacuum (in order to enable the field source to emit energy permanently), and in both cases the propagating field gives back energy to the vacuum.

By the way it can be mentioned, that the energy and the power, which the field sources extracts from the vacuum (in the electrostatic case as well as in the magnetic case) is larger than the energy and the power, which the vacuum takes back during the propagation of the field. This is plausible, because the volume filled with field permanently grows during time (without a decrease of the field strength at a given position). The increase of the field strength during time has the consequence that the vacuum permanently loses some energy to the field.

Consequence: In analogy to the electric field and the circulation of energy herein, the circulation of energy of the magnetic field should also offer a possibility to extract energy from the vacuum. And both types of field- and energy- circulation should be observable in laboratory, if a mechanism can be found, which allows to extract energy and tu use it for the drive of a mechanical device (for instance to make a rotor rotate).

3. Theoretical fundament of the energy-flux

Of course, there is a connection between the energy-flux described in section 2 and the energy density of the vacuum.The energy density of the vacuum is still nowadays entitled as one of the unsolved puzzles of physics [Giu 00]. At least a certain part of the vacuum energy should be explainable by a summation of the eigen-values of the energy of the zero point oscillations of the vacuum (of electromagnetic harmonic oscillators resp. waves) [Whe 68]. Nevertheless, this statement does not want to express, that this special energy sum is the only energy within the vacuum. But this energy-sum can seen in relation with the energy-flux described in section 2. This will one of the topics of section 3.

The problem with the summation of the eigen-values of the energy of all zero point oscillations (these are infinitely many) is the divergence of the improper integral over all wave vectors of these zero point oscillations. One approach to the solution is discussed in Geometrodynamics, which is nowadays seen with large scepticism because it is in contradiction with measurements of Astrophysics (see section 3.5).

In section 3 of the present work a new solution for this convergence problem is introduced on the basis of Quantum electrodynamics (where improper integrals are not unknown), and this solution comes to values appearing realistic [e6]. The only necessary postulate is: It is well known that the speed of propagation of electromagnetic waves (in the vacuum) is influenced by electric and magnetic DC-fields [e7], [Eul 35], [Rik 00] [Bia 70], [Boe 02], [Ost 07]. **The postulate is now to assume, that the zero point oscillations of the vacuum display the same behaviour as the other waves.**

Based on this concept we will now investigate the energy of the zero point oscillations of the vacuum, and we will find an idea how this energy can be made manifest in the laboratory. The experiments to realize this idea will be presented later in section 4.

3.1. Vacuum-energy in Quantum mechanics

As generally known in Quantum theory, the eigen-values of the energy of electromagnetic waves (they do a harmonic oscillation) are given as $\left(n+\frac{1}{2}\right)\hbar\omega$, where n is number of photons, calculated as the eigen-value of $\hat{a}^{\dagger}\hat{a}$ with regard to the wave function ψ_n in the equation $\hat{a}^{\dagger}\hat{a}\psi_n = n\cdot\psi_n$ (with $\hat{a}^{\dagger}$ being the operator of particle creation and $\hat{a}$ being the operator of particle annihilation). As long as no particles are present, we have $n=0$, and the eigen-values of the energy of the ideal (physical) vacuum $|0\rangle$ (in Dirac's notation) are found by integration over all frequencies ω respectively over all wave vectors $\vec{k}$ in the k-space leading to

$$E = \int \frac{1}{2} \hbar \omega \, d^3\vec{k} \quad (1.28)$$

(without consideration of polarization) [Man 93].

We know that this integral is divergent, because for small wavelength $\lambda \to 0$ (which fit well into every small volume), the absolute values of the wave vector $|\vec{k}| = \sqrt{k_x^2 + k_y^2 + k_z^2}$ as well as the frequency ω go to infinity. This brings the problem that the integral in (1.28) leads to an infinite energy density. Normally this energy density is treated as a constant without significance to physics, which is eliminated by setting the zero point of energy to the ground state $|0\rangle$ of the vacuum [Kuh 95] (on top of all the harmonic oscillations). The creation of a photon $\hat{a}^{\dagger}|0\rangle = |1\rangle$ leads to the excited state of a harmonic oscillation, of which the energy eigen-value is for $1 \cdot \hbar\omega_{\vec{k}} = \hbar c |\vec{k}|$ above the energy of the ground state [Köp 97].

The propagation of a photon in the vacuum without a field (electric and magnetic) follows the speed of light. Because the propagation of electrostatic and magnetic fields are understood as the exchange of photons [Hil 96], the logical consequence should be, that electric and magnetic DC-fields should also follow this speed of propagation. Assigning this conception to DC-fields is not usual for everybody, but we will find further justification in the following chapters with arguments within the Theory of Relativity and with arguments within Quantum Theory. The background will be understandable within the model presented in section 3 of the present work.

The fact, that even the ground state $|0\rangle$ of the "empty vacuum" contains the energy of harmonic oscillations of electromagnetic waves, is the reason that they got the name "zero point oscillations". Their energy according to (1.28) defines the vacuum-energy of the ground state. If we want to have access to this energy (and to convert it into a classical form of energy), we have to understand their nature. A well-known example for a force coming out of this type of vacuum-energy is the Casimir-force [Cas 48], [Moh 98], [Bre 02], [Sve 00], [Ede 00], [Lam 97], which is explained on the basis of zero point oscillations. This explanation is based on the analysis of the influence of two ideally conducting (metallic) plates onto the spectrum of zero point oscillations of the vacuum. The free vacuum (i.e. without those plates) consists of a continuous spectrum of all imaginable wavelengths, whereas the space between the plates only contains a discrete spectrum of resonant (standing) waves, because the plates act as reflectors with a field strength of zero at the surface of each plate, defining nodal points of the oscillation. From the energy-difference between those both spectra (in the free vacuum and between the plates), Casimir deduces the energy density and the force between the plates.

This arises the expectation that the conversion of vacuum-energy into mechanical energy as reported in [e8] can be understood in analogy to the Casimir-effect. If this energy conversion shall be done in a perpetual process, we have to find a possibility to move the plates relatively to each other without alteration of the distance between them. This has been developed in the present work. Thus, there is some similarity with the Casimir-effect, but there is also an important difference: If some of the energy of the zero point oscillation shall be converted into mechanical energy, the conducting plates have to move relatively to each other, but they are not allowed to alter their distance (which leads us to a parallel shift) – otherweise the conversion would not be perpetual. This is imaginable if the plates perform an appropriate rotation. The practical setup is presented in section 4.

Although the Casimir-effect helped to invent a machine which verifies the existence of the energy of the zero point oscillations and converts it into mechanical energy, it was clear from the very beginning of the development, that the metal plates (whose position have to be different from the position in the Casimir configuration) and the vacuum, necessary for the Casimir-effect are not enough for the endless conversion of vacuum-energy. Additionally to those objects, there has to be an electric (or a magnetic) field, which has to provide the possibility to interact with the energy circulation of section 2.

The experiments which finally succeeded in converting vacuum-energy into classical mechanical energy [e8, e9] confirm this approach. The first explanation of the functioning principle of the energy-converting rotor was given on the basis of an electrostatic field within classical electrodynamics [e5], [e9], [e10]. The logical connection between the zero point oscillations and classical electrodynamics is topic of [e6]. A central aspect thereby is the mechanism how the electric and the magnetic fields propagate in the vacuum (i.e. into the space). The crucial question is: How do those fields and their propagation influence the zero point oscillations. This question will also be answered in the following chapters.

3.2. Connection with the classical model of vacuum-energy

Before we answer the crucial concluding question of section 3.1 (the interaction between the propagating field and the zero point oscillations), we want to recapitulate the basics of the model of section 2, which already showed the way how to convert vacuum-energy into classical energy:

Electric and magnetic fields as regarded in Classical Electrodynamics are normally regarded to be "everywhere in space at the same moment" [Kli 03]. This means, that normally their time dependent propagation into the space is not taken into consideration, but only their presence. For most of the technical and practical applications of electrodynamics (with typical distances inside the laboratory and velocities negligible in comparison with the speed of light) this is fully sufficient. But as a matter of principle, this is in clear contradiction with the Theory of Relativity [Goe 96], according to which the propagation of the field strength has to respect at least the limit of the speed of light [Chu 99]. Thus, it appears sensible to take the propagation (of DC-fields) with the speed of light into account. But this conception leads to important consequences – one of them is the awareness, that electric and magnetic fields give energy to the vacuum during their propagation (as stated in section 2).

Even though this fundamental logic of Classical Electrodynamics is sufficient to explain the existence and the nature of the energy circulation in the vacuum, we want to look at the inner structure of the vacuum in order to find the backgrounds of the described behaviour. In order to prepare the microscopic model of energy conversion, we want to outline the circulation of the energy in the vacuum:

The propagation of an electric field as well as a magnetic field (in the model to be developed now) influences the wavelength of the zero point oscillations. We will find the correlation between the field strength and the alteration of the wavelength soon. The central assumption of the model is, that quantum electrodynamical corrections such as vacuum polarisation do not only occur with photons but also with zero point oscillations (and we will find, that this causes the extraction of energy from the field). But the particles of vacuum polarisation do not follow the propagation of the field, and so they can distribute their energy all over the space. This is the "drain" into which the field loses its energy during its propagation. On the other hand, this mechanism also gives us the explanation of the source, from which the electrical charges are permanently supplied with energy in order to produce field strength: It indicates the reason for the transportation of field energy to be an alteration of the wavelength of the zero point oscillations. And there is further conclusion, that the lost field energy during the propagation of the field is the energy necessary for vacuum polarization.

In the following sections 3.3 and 3.4 we will see, that this model does not only explain the experiments of the conversion of vacuum-energy but it furthermore also allows to determine the energy density of the zero point oscillations in the vacuum.

3.3. New microscopic model for the electromagnetic part of the vacuum-energy

Annotation: It should be mentioned, that the present work only analyzes the connections between the field energy of the electric and the magnetic field on the one hand and its part of the vacuum-energy on the other hand. The present work does not give any answer to the question whether there are some further other items in the vacuum, giving further contributions to the vacuum-energy, which we do not know today. In the moment, mankind does not have an imagination how to answer this question, and it is not topic of the present work.

Clear in any case is, that the zero point oscillations mentioned above can be understood in connection with several effects of vacuumpolarization (such as for instance virtual electrons and positrons) [Fey 97], [Gia 00]. Thus, the energy of the zero point oscillations should be explainable with such physical items. With other words: An explanation has to be searched which helps to understand the propagation of electric and magnetic fields, as well as the supply of field sources with energy, on the basis of the items of the vacuum.

The model for this explanation has been found in the present work in such simple (and elementary) manner, that it is advantageous according to Occam's razor [Sim 04], which always prefers explanations as easy as possible. Our model finally goes back to the year 1935, when Heisenberg and Euler [Hei 36] performed quantum theoretical calculations of the Lagrangeian of photons in electric and magnetic fields, coming to the conclusion that photons propagate in such fields with lower speed than in the vacuum without field. The reasons are found in vacuum polarisation, which influences the Lagrangeian as calculated by Heisenberg and Euler. The experimental verification is not yet completely done. It was regarded as complete in [Zav 06], but this scientist later withdrew his results [Zav 07] by himself. But it is supposed that the verification will be done in not too far future [Che 06], [Lam 07], [Bes 07].

Logical consequence leads to a conclusion, which is the only assumption of our model:

If electromagnetic waves (such as photons) undergo retarded propagation within electric and magnetic fields (in comparison with field free vacuum), zero point oscillations should undergo the same retardation of propagation, because they have the same nature (to be electromagnetic waves). This means that electric and magnetic fields have an influence on the wave vectors $\vec{k}$ and on the frequencies ω of the zero point oscillations. This causes an influence on the energy eigen-values of the zero point oscillations. This postulate is one of the main fundamental considerations of the present work. We want to use it as a hypothesis (and the experiment will confirm it later):

The alteration of the energy of the zero point oscillations in electric and magnetic fields should be sufficient to explain the energy of those fields.

The development of out QED-model can now be done in similarity with Casimir's considerations (and the effect with his name) by comparing the energy of the continuous spectrum of the zero point oscillations with and without intervention. Casimir's intervention was to mount two conducting plates (without electrical charge). The intervention of the present work is to switch on an electric (or a magnetic) field between some nonparallel plates. The result, which we have to find, is the difference of the total energy of the spectra with and without field. The difference of those both total energy sums should be directly the energy of the field. In principle, our calculation has the same problems with the convergence of improper integrals as Casimir's calculation. And the similar problems should be solvable in a similar way. The typical method for the solution is renormalization in Quantum field theory [She 01], [She

03], see also [Hoo 72], [Dow 78], [Bla 91]. The mathematical methods are presented also very clearly in [Kle 08]. But the result can be obtained most easy, if we can find and use utilisable results somewhere in literature as done in the following calculation:

The energy density of the electromagnetic field, resp. of its zero point oscillations is calculated oftentimes in the $\vec{k}$-space [She 01], making use of $\vec{p}=\hbar\cdot\vec{k}$ as leading to the commonly known equation (1.29)

$$\left.\frac{E}{V}\right|_Z = s\cdot\int E_0(\vec{k})\frac{d^3k}{(2\pi)^3}\,, \tag{1.29}$$

where $E_0(\vec{k})$ is the spectrum of the energy of the zero point oscillations, so that the integration is going continuously over all possible $\vec{k}$-vectors. The index "Z" stands for "zero point oscillations". The vacuum-energy is related to the transition-amplitude $\langle 0|0\rangle$ from the vacuum to the vacuum, which is represented by close loops for virtual particles in Feynman diagrams. The "s" is 1 because we want to consider the different states of polarisation separately. They correspond to the well known energy eigen-values of $E_0(\vec{k})=(n+\frac{1}{2})\hbar\omega$, where $n=0$ is the ground state. So we have to insert $E_0(\vec{k})=\frac{1}{2}\hbar\omega$ into (1.29) and we receive

$$\left.\frac{E}{V}\right|_Z = \int\frac{1}{2}\hbar\omega\frac{d^3k}{(2\pi)^3}\,. \tag{1.30}$$

Furthermore, the isotropy of the space allows us to work with the absolute values of the $\vec{k}$-vector and to write $\omega=c\cdot|\vec{k}|$, or in Cartesian coordinates $\omega=c\cdot\sqrt{k_x^2+k_y^2+k_z^2}$. Thus, we come from (1.30) to (1.31):

$$\left.\frac{E}{V}\right|_Z = \frac{1}{2}\cdot\int\hbar c\cdot\sqrt{k_x^2+k_y^2+k_z^2}\;\frac{dk_x\cdot dk_y\cdot dk_z}{(2\pi)^3} = \frac{1}{2}\hbar c\cdot\int|\vec{k}|\frac{d^3k}{(2\pi)^3}\,. \tag{1.31}$$

The divergence of this improper integral is commonly known, because the wave vector $\vec{k}$ goes to infinity for small wavelength – and all these wave vectors have to be taken into account in the improper integral.

In analogy to Casimir's thoughts, we are not mainly interested (as explained above) in the limes of this integral, but we are mainly interested in the difference of the limes of this integral for $\vec{k}$-vectors with and without field. This means, we want to know the limes of

$$\left.\frac{E}{V}\right|_{FIELD} = \left.\frac{E}{V}\right|_{Z,WITH} - \left.\frac{E}{V}\right|_{Z,WITHOUT} = \left(\frac{1}{2}\hbar c \iiint \left|\vec{k}_{Z,WITH}\right| \frac{d^3k}{(2\pi)^3}\right) - \left(\frac{1}{2}\hbar c \iiint \left|\vec{k}_{Z,WITHOUT}\right| \frac{d^3k}{(2\pi)^3}\right), \tag{1.32}$$

where the indices "WITH" and "WITHOUT" represent the situation with and without external field. This difference must be the energy density of the field, as marked by the index "FIELD".

And this is the criterion of evaluation for our model:

The model must allow the calculation of the energy density of the zero point oscillations of the vacuum from the field-energy $\left.\frac{E}{V}\right|_Z$ for both types of fields (the electric as well as the magnetic), and the energy density of the vacuum must be the same for both types of fields. This means that the alteration of the energy of the zero point oscillations caused by the $\vec{k}_{Z,WITH}$-vectors of the electric field has to be the same as the alteration of the energy of the zero point oscillations caused by the $\vec{k}_{Z,WITH}$-vectors of the magnetic field. Because the energy density must not depend on the way of calculation, the results of both ways have to lead to the same energy density $\left.\frac{E}{V}\right|_Z$. Only if this condition is fulfilled, our model is sensible.

Under this view, we regard electric and magnetic fields to be two different probes for the analysis of the energy density of the vacuum.

We prepare this evaluation with an introductory remark, applicable for both ways of calculation, before we will come to each way of calculation in a separate consideration. [Boe 07] gives the Heisenberg-Euler-Lagrangeian from [Hei 36] in SI-units (for the sake of understandability):

$$\begin{aligned} L &= -\frac{c^2\varepsilon_0}{4} F_{\mu\nu}F^{\mu\nu} + \frac{\alpha^2\hbar^3\varepsilon_0^2}{90\, m_e^4 c}\left[\left(F_{\mu\nu}F^{\mu\nu}\right)^2 + \frac{7}{4}\left(\tilde{F}_{\mu\nu}\tilde{F}^{\mu\nu}\right)^2\right] \\ &= \frac{\varepsilon_0}{2}\left(\vec{E}^2 - c^2\vec{B}^2\right) + \frac{2\alpha^2\hbar^3\varepsilon_0^2}{45\, m_e^4 c^5}\left[\left(\vec{E}^2 - c^2\vec{B}^2\right)^2 + 7c^2\left(\vec{E}\cdot\vec{B}\right)^2\right], \end{aligned} \tag{1.33}$$

where m_e is the mass of the electron and the other symbols are to be interpreted as usually.

There are several articles in which the speed of propagation of electromagnetic waves in electric, magnetic and electromagnetic DC-fields are calculated on the basis of this Heisenberg-Euler-Lagrangeian (see for instance [Lam 07], [Hec 05], [Lig 03], [Rik 00], [Rik 03], [Riz 07], [Sch 07], [Zav 07]). And we want to assign this speed of propagation also to the electromagnetic waves of the zero point oscillations of the vacuum, thus we will take the speed of propagation from those articles in order to determine the influence of the external field (the electric as well as the magnetic) onto the $\vec{k}$-vector of the zero point oscillations and thereby the influence on energy density of the zero point oscillations of the vacuum, which is $\left.\frac{E}{V}\right|_Z$. These results contain the solution of the improper integrals mentioned above.

We will do this now for both probes (types of fields interacting with the zero point oscillations, i.e. the electric and the magnetic field) separately, beginning with the magnetic field (first calculation) and later followed with the electric field (second calculation).

1. calculation → For the determination of $\left.\frac{E}{V}\right|_Z$ with a magnetic field as a probe:

[Boe 07] gives the influence of a magnetic field on the speed of propagation of electromagnetic waves as

$$1-\frac{v}{c}=a\cdot\frac{\alpha^2\hbar^3\varepsilon_0}{45m_e^4c^3}\cdot\left|\vec{B}\right|^2\cdot\sin^2(\theta)=\begin{cases}5.30\cdot10^{-24}\frac{1}{T^2}\cdot\left|\vec{B}\right|^2\cdot\sin^2(\theta) & \text{für } a=8,\Box\text{–Modus}\\ 9.27\cdot10^{-24}\frac{1}{T^2}\cdot\left|\vec{B}\right|^2\cdot\sin^2(\theta) & \text{für } a=14,\perp\text{–Modus}\end{cases} \qquad (1.34)$$

$$\left(\text{with } \left|\vec{B}\right| \text{ in Tesla}\right),$$

where the direction of the propagation of the photon and the direction of the magnetic field enclose an angle of θ, and they define a plane, which is the reference to classify the $\Box$-mode ($a=8$) and the $\perp$-mode ($a=14$) of polarization. We have two different speeds of propagation, namely v with external field and c without external field. The difference of both speeds leads us to the Cotton-Mouton-birefringence of the vacuum, being

$$\Delta n_{Cotton-Mouton}=\left(1-\frac{v}{c}\right)_{\perp}-\left(1-\frac{v}{c}\right)_{\Box}=3.97\cdot10^{-24}\frac{1}{T^2}\cdot\left|\vec{B}\right|^2\cdot\sin^2(\theta)\ , \qquad (1.35)$$

which is confirmed by [Rik 00] quantitatively for an angle of $\theta=90°$ and also by [Bia 70]. The last-mentioned reference is often regarded as a milestone on the way to the comprehension of electromagnetic waves in electric and magnetic DC-fields, because it is the first work giving quantitative predictions of the birefringence (and thus of the speed of propagation) of electromagnetic waves in the fields, which is a quantity that can be really measured.

The connection between the frequency ω and the speed of propagation of electromagnetic waves can be concluded from the fact, that the range of the integration over the $\vec{k}$-vectors is the same with and without the field. Thus $\omega = c \cdot |\vec{k}|$ leads to the consequence that $|\vec{k}| = \frac{\omega}{c}$, where the absolute values of the wave vector is independent of the question whether a field is applied or not. So we have the relation of the

$$\frac{\omega_{WITHOUT}}{c} = \frac{\omega_{WITH}}{v} \quad \Rightarrow \quad \omega_{WITH} \cdot c = \omega_{WITHOUT} \cdot v \quad \Rightarrow \quad \omega_{WITH} = \omega_{WITHOUT} \cdot \frac{v}{c}, \tag{1.36}$$

with c as the speed of propagation without field and v as the speed of propagation with field.

The energy density of the magnetic field is known from classical electrodynamics [Jac 81] to be

$$\left.\frac{E}{V}\right|_{FIELD} = \frac{\mu_0}{2} \cdot \vec{H}^2 = \frac{1}{2\mu_0} \cdot \vec{B}^2 \ . \tag{1.37}$$

Equations (1.30) and (1.32) together with (1.36) lead to the expression

$$\begin{aligned} \frac{1}{2\mu_0} \cdot \left|\vec{B}\right|^2 = \left.\frac{E}{V}\right|_{FIELD} &= \int \frac{1}{2}\hbar \cdot \omega_{Z,WITHOUT} \frac{d^3k}{(2\pi)^3} - \int \frac{1}{2}\hbar \cdot \omega_{Z,WITH} \frac{d^3k}{(2\pi)^3} \\ &= \int \frac{1}{2}\hbar \cdot \omega_{Z,WITHOUT} \frac{d^3k}{(2\pi)^3} - \int \frac{1}{2}\hbar \cdot \omega_{Z,WITHOUT} \cdot \frac{v}{c} \frac{d^3k}{(2\pi)^3} . \end{aligned} \tag{1.38}$$

$$\text{From (1.35)} \quad \Rightarrow \quad \frac{1}{2\mu_0} \cdot \left|\vec{B}\right|^2 = \left.\frac{E}{V}\right|_{FIELD} = \left(1 - \frac{v}{c}\right) \cdot \left(\int \frac{1}{2}\hbar \cdot \omega_{Z,WITHOUT} \frac{d^3k}{(2\pi)^3} \right), \tag{1.39}$$

because the external fields alter the frequency and the energy of each quantum mechanical zero point oscillation, according to our model.

With (1.34) we find the sum of the energy of all zero point oscillations, because the field strength of the magnetic field is dispensed from the equation:

$$\frac{1}{2\mu_0}\cdot\left|\vec{B}\right|^2=\left(1-\frac{v}{c}\right)\cdot\int\frac{1}{2}\hbar\cdot\omega_{Z,WITHOUT}\frac{d^3k}{(2\pi)^3}$$

$$=a\cdot\frac{\alpha^2\hbar^3\varepsilon_0}{45\,m_e^4c^3}\cdot\left|\vec{B}\right|^2\cdot\sin^2(\theta)\cdot\int\frac{1}{2}\hbar\cdot\omega_{Z,WITHOUT}\frac{d^3k}{(2\pi)^3} \tag{1.40}$$

$$\Rightarrow\quad\int\frac{1}{2}\hbar\cdot\omega_{Z,WITHOUT}\frac{d^3k}{(2\pi)^3}=\frac{1}{2\mu_0}\cdot\frac{1}{a}\cdot\frac{45\,m_e^4c^3}{\alpha^2\hbar^3\varepsilon_0}=\frac{1}{a}\cdot\frac{45\,m_e^4c^5}{2\cdot\alpha^2\hbar^3}=\frac{1}{a}\cdot 6.007\cdot 10^{30}\frac{J}{m^3}, \tag{1.41}$$

at an excitation with $\theta=90^\circ$, where m_e is the mass of the electron and α is the constant of hyperfine structure.

This is the way how all convergence problems of improper integrals over spectra of zero point oscillations have been traced back to existing solutions in literature.

Free from any problems of convergence of improper integrals, and free from any cut-off functions (with uncertain motivation), we found in literature, some results, that help to calculate the energy density of the zero point oscillations of the vacuum. We have in mind that the calculation was done with a magnetic field as a probe, but the influence of the magnetic field was eliminated during the course of the calculation.

It can also be seen, that the $\perp$-modes and the □-modes can be excited with different strength, but this is also an aspect of the magnetic field as a probe. And it must not influence the answer to the fundamental question about the energy density of the zero point oscillations of the vacuum. In the same way, the choice of the probe (magnetic or electric field) to interact with the zero point oscillations must not influence the energy density of the electromagnetic zero point oscillations of the vacuum. We have indeed to keep in mind, that different probes can excite oscillations differently, so we have to extract a measurable quantity from our result, which will allow a comparison with the result of the second way of calculation, which is done with an electric field as a probe.

Such a quantity is the birefringence of the vacuum, which several articles regard as the central quantity of measurement. This shall be determined now in order to create a possibility to compare the result of the first calculation with the result that will follow from the second calculation.

This measurable quantity is the birefringence of the vacuum, which several articles regard as the central quantity of measurement (see above: [Lam 07], [Lig 03], [Rik 00], [Rik 03], [Riz 07], [Sch 07], [Zav 07]).

We achieve the comparability as following:

Also the difference, which represents the measurable birefringence, must lead to the same energy density of the vacuum, this is

$$a_{\perp}\cdot\left[\int\frac{1}{2}\hbar\cdot\omega_{Z,WITHOUT}\frac{d^3k}{(2\pi)^3}\right]-a_{\parallel}\cdot\left[\int\frac{1}{2}\hbar\cdot\omega_{Z,WITHOUT}\frac{d^3k}{(2\pi)^3}\right]=\frac{45m_e^4c^3}{\alpha^2\hbar^3\varepsilon_0} \tag{1.42}$$

$$\Rightarrow\quad(14-8)\cdot\left[\int\frac{1}{2}\hbar\cdot\omega_{Z,WITHOUT}\frac{d^3k}{(2\pi)^3}\right]_{\perp-\parallel}=\frac{45m_e^4c^5}{2\cdot\alpha^2\hbar^3}=6.007\cdot10^{29}\frac{J}{m^3} \tag{1.43}$$

$$\Rightarrow\quad\left[\int\frac{1}{2}\hbar\cdot\omega_{Z,WITHOUT}\frac{d^3k}{(2\pi)^3}\right]_{\perp-\parallel}=\frac{1}{a_{\perp}-a_{\parallel}}\cdot\frac{45m_e^4c^5}{2\cdot\alpha^2\hbar^3}=\frac{1}{14-8}\cdot6.007\cdot10^{29}\frac{J}{m^3}$$
$$=1.001\cdot10^{29}\frac{J}{m^3} \tag{1.44}$$

This measurable value, which results from fundamental considerations of the birefringence of electromagnetic waves in magnetic fields has to be kept it in mind for later comparison with the birefringence of electromagnetic waves in electric fields, which will be the result of the second way of calculation.

2. calculation → For the determination of $\left.\frac{E}{V}\right|_Z$ with an electric field as a probe:

According to [RIK 00] the Kerr-birefringence of electromagnetic waves in electric fields has the absolute value of

$$\Delta n_{Kerr}\approx4.2\cdot10^{-41}\frac{m^2}{V^2}\cdot\left|\vec{E}\right|^2 \tag{1.45}$$

(with the electric field strength in V/m), with a possible uncertainty in the last digit due to truncation of the numerical value. The value is also confirmed by [Bia 70].

From there we can easily find the absolute value of the energy density difference in analogy with (1.44) as following

$$\frac{\varepsilon_0^2}{2}\left|\vec{E}\right|^2 = \Delta n_{Kerr} \cdot \left[\int \frac{1}{2}\hbar \cdot \omega_{Z,WITHOUT} \frac{d^3k}{(2\pi)^3} \right] \tag{1.46}$$

$$\Rightarrow \quad \int \frac{1}{2}\hbar \cdot \omega_{Z,WITHOUT} \frac{d^3k}{(2\pi)^3} \approx \frac{\frac{\varepsilon_0^2}{2}\left|\vec{E}\right|^2}{-4.2 \cdot 10^{-41} \frac{m^2}{V^2} \cdot \left|\vec{E}\right|^2} \approx 1.0 \cdot 10^{29} \frac{J}{m^3} \tag{1.47}$$

This is the energy density difference for birefringence, determined with an electric field as a probe. Of course, the field strength as an attribute of the probe had to be eliminated, so that the result is free from any information about the probe. Thus, both calculations no.1 and no.2 should come to the same result, because there is only one vacuum, for which both calculations are valid. The fact, that the solutions fulfil this criterion, demonstrates that we have indeed found a possibility to trace back the convergence problems of (1.29), (1.30) and (1.31) to other results found in literature. And it confirms our model of the propagation of electric and magnetic DC-fields with the model assumptions as described above.

For the sake of completeness, it should be mentioned again, that the calculation only gives the energy density of the electromagnetic zero point oscillations of the vacuum. This is not a general value for the total energy density of the vacuum. It does not contain any considerations of fundamental interactions other than the electromagnetic interaction and their mechanisms. This is said in order to remember, that we have been thinking only about the electromagnetic part of the vacuum-energy. Perhaps other fundamental interactions also have some contribution to the total vacuum-energy and might allow also the conversion of their parts of the vacuum-energy (into a classical form of energy) one day.

3.4. The energy-flux of electric and magnetic fields in the vacuum, regarded from the view of QED's zero point oscillations

Up to now, the model, which was introduced in section 3, is capable to explain the propagation of electric and the magnetic DC-fields in the vacuum. The only assumption it needs is surprisingly simple and plausible:

From several literature references it is known, that the propagation of excited states $|n\rangle$ of electromagnetic waves (for $n \geq 1$) in the vacuum is influenced by electric and magnetic fields. The assumption of our model is, to apply the same dependence on electric and magnetic fields also to the propagation of the ground state (for $n = 0$).

This assumption looks logic and plausible. One of the consequences is, that the propagation of electric and magnetic DC-fields can be understood as an alteration of the wavelengths and frequencies of the ground state (these are the zero point oscillations).

But there are further aims of the present work, one of them is: We want to find a possibility to make the vacuum-energy manifest in the laboratory. There we need an explanation of the above mentioned energy-flux of the electric and the magnetic field and the propagation of these fields. As soon as this energy-flux is understandable, we can plan an experiment for the conversion of the vacuum-energy into mechanical energy.

The explanation of the energy-flux can be given as following:

Let us begin with an explanation of the propagation of the fields, this is an answer to the question by which means the vacuum permanently takes energy out of the field during the propagation of the field – and an answer to the question by which means the vacuum permanently supplies the field source with energy. This can be explained as following:

The quantum electrodynamical corrections in the Heisenberg-Euler-Lagrangeian correspond to some certain energy (otherwise they would not be seen in the Lagrangeian). And each energy-term corresponds to a certain event of vacuum polarisation. This is the case for those energy-terms which Heisenberg and Euler took into account same as for higher order energy-terms corresponding to several higher-order effects of vacuum polarisation. The zero point oscillations give their contributions to all these energy-terms (otherwise, they would not be influenced by the fields), and the consequence is, that the field gives contributions to these energy-terms. Even if the events of vacuum polarisation are only temporarily in the time they occur, they permanently repeat with their certain amplitudes of probability [Fey 85], [Fey97], [Fey 49a], [Fey 49b], [Sch 49]. This describes a dynamic equilibrium situation, where a given number of several effects of vacuum polarisation happen within a given interval of time – always handling some energy within the vacuum. And the number of effects of vacuum polarisation happening within a given interval of time is influenced by an external field. The situation is a process of dynamical equilibrium, which can be shifted by an applied (electric or magnetic) field.

With other words: In our model, the field generates a lot of events of vacuum polarisation during its propagation (because of the field's energy) – and this is just the amount of energy, which the field gives to the vacuum during its propagation. And this is necessary in order to enable the field strength to follow Coulomb' law. And the energy necessary for the alteration of the number of vacuum polarisation events is the extracted energy from the field.

The way how the field source is supplied with energy from the vacuum follows the same principle but to the opposite direction:

The events of vacuum polarisation take finite time and finite space, but they don't have to follow the directions of the flux lines of the field. Thus they diffuse (perhaps statistically) to every direction and to everywhere in the space. Obviously they generate the loss of energy from the field during its propagation, which forms the observed energy-flux (see (1.12), (1.13), (1.14), (1.25) for electric fields and (1.18), (1.22), (1.23), (1.24) for magnetic fields), which transports energy through the vacuum and which also can be tapped by field sources. This means that every field source is permanently supplied from this energy circulating around in the vacuum.

It is not obligatory, that an electrical charge is supplied with energy from its own field. (It can be supplied with energy from the field of some other charges as well.) But it is obligatory, that each electrical charge permanently takes more energy from the vacuum than its field gives to the vacuum. The reason is simple: The amount of volume filled with field grows permanently during time, and with it the field's total energy (because the field strength remains constant at a given position). One of the consequences of this fact is, that growing total energy will cause a growing number of events of vacuum polarisation during time. And: The total field's energy of all electric (and magnetic) fields in the universe permanently increases during time. (This might probably have a consequence for cosmology, perhaps for the expansion rate of the universe.)

An explanation, which is not available in the moment, concerns the mechanism, by which the field source is able to convert the energy of the events of vacuum polarization into field energy. But such a background explanation is not necessary for the verification of the vacuum-energy as described in the experiment in section 4.

In the notation of particle physics, the zero point oscillations are bosonic Quantum field fluctuations (because they are electromagnetic waves), and events of vacuum polarisation are fermionic Quantum field fluctuations (because they consist of particles like virtual electrons and positrons) [She 01], [She 03]. Their conversion (which is necessary for vacuum polarisation) consists of processes like virtual pair production and annihilation. Especially in electric and magnetic fields, the probability amplitudes for the conversion of those both types of Quantum field fluctuations depend on external field strength. Thus the distance to the field source has an influence on the number of such processes occurring in a given time interval.

A statement regarding the energy loss of a field during its propagation (which should be connected with the probability amplitudes of the events of vacuum polarisation) can be given on the basis of a simple classical consideration following (1.12) and (1.13) and (1.15) for the example of a spherical shell (around a point charge) filled with electric field. When this field-volume propagates from $x_1 \ldots x_1 + c \cdot \Delta t$ to $x_2 \ldots x_2 + c \cdot \Delta t$, it releases the energy $\Delta E = E_{\substack{inner \\ shell}} - E_{\substack{outer \\ shell}} \approx \frac{Q^2}{8\pi\varepsilon_0} \cdot \frac{2 \cdot c \cdot \Delta t \cdot \Delta x}{x_1^3}$.

Especially the energy going from the field to the vacuum follows the condition: The number of events of vacuum polarisation (connected with the probability amplitudes) is responsible for the amount of energy being converted from field-energy to vacuum-energy.

After all, we face the question, how an experiment should be designed in order to convert vacuum-energy into mechanical energy. In section 4 we will see, that one part of the experiment is a rotor, but how shall this rotor be built and what else do we need to get a machine, which extracts energy from the circulation of field's energy and vacuum-energy ?

Of course, the answer has to be based on the field's energy, because the energy converting movement of the rotor is finally driven by electrostatic forces (or in the case of the magnetic rotor by magnetostatic forces). In our model, some metallic rotor blades do interact with the circulation of energy, namely as following:

That part of the circulating energy, which is contained in the field strength, alters the wavelengths of the zero point oscillations. This is evoked by the field source, because the field source produces the field. As known from the Casimir-effect, conductor plates block the field and with it the propagation of the zero point oscillations. The consequence is, that the wavelengths of the zero point oscillations are influenced by the field on this side of the conductor plate, which looks towards the field source, but on the opposite side (looking away from the field source), the field does not have any influence on the wavelengths of the zero point oscillations. This means, that the wavelengths of the zero point oscillations are different on both sides of the conductor plates, which is only possible with regard to energy conservation, if the conductor plates compensate the difference of the energy. As soon as the conductor plates absorb energy, they feel a force, which drives the rotor described in section 4. And the fact that the conductor plates really absorb energy from the field (and do not emit energy into the field) is clear, because at their side looking away from the field source, there is no field and thus no field's energy. This means that the conductor plates prevent some space from getting field and field's energy.

With other words: The conductor plates absorb exactly this amount of energy, which can not propagate into the space behind the plates. Finally, this absorbed energy is being transported with the energy-flux from the field source to the plates. And this is the energy, from which we have to prove (in section 4) that it drives the rotor by the following mechanism:

The field source converts vacuum-energy into field energy by altering the wavelengths of the zero point oscillations. This alteration is stopped abruptly at the surface of the conductor plates (these are the blades of the rotor), so that the conductor plates absorb some field energy from the flux of the propagating field. (They absorb the percentage of the flux, which hits their surface.) Nevertheless, it is much easier to calculate the forces driving the vacuum-energy rotor if we follow the formalism of Classical Electrodynamics, using the image charge method [Bec 73], as will be done soon when developing the real concept of the experiment for the verification of the vacuum-energy and its conversion into mechanical energy.

3.5. Comparision of the QED-model with other models

We now intend to compare our value for the energy-density of the vacuum (see (1.44) and (1.47)) with the values of other models in the context of other disciplines of physics. But we have to keep in mind, that our model only gives the energy-density of the electromagnetic zero point oscillations, not the total energy-density of the vacuum.

The total energy-density of the vacuum is a topic of Cosmology, because it refers to the gravitation, caused by the mass of the vacuum. The reason is that every type of energy corresponds to a mass, which (finally) gives rise to gravitation. And this is also the case for the vacuum-energy contained in the universe, which influences the expansion of the universe because of its gravitation. From the alteration of the speed of the expansion of the universe, it is possible to derive the total mass of the vacuum (together will all other matter within the universe). Thus the cosmological value gives the total energy-density of the vacuum, which should be larger than the value of our model. But this will lead to an open question, which can be seen as a contradiction (see (1.48) below).

The question of the energy density of the universe is known to be one of the unsolved puzzles in nowadays physics, and it is often called "the largest discrepancy known in Physics" with a range of more then 120 orders of magnitude between different values (for the same energy-density) from different branches on physics.

On the one hand we can find many publications of cosmology (see for instance [e1], [GIU 00], [TEG 02], [EFS 02], [TON 03], [RIE 98]), giving values of the density of mass ρ_M and/or the energy density ρ_{grav} of the universe based on considerations of gravitation in connection with the expansion of the universe. The average of these values come to a range of

$$\rho_M \approx (1.0 \pm 0.3) \cdot 10^{-26} \frac{kg}{m^3} \quad \Rightarrow \quad \rho_{grav} = c^2 \cdot \rho_M = (9.0 \pm 0.27) \cdot 10^{-10} \frac{J}{m^3} \qquad (1.48)$$

The fundaments of the expansion of the universe are experimental observations.

On the other hand there are values coming from Geometrodynamics [WHE 68], which are determined by pure theoretical considerations, based on the zero point oscillations of the vacuum plus the hypothesis that such oscillations can not exist with a wavelength below the Planck-scale. This additional assumption is just made to get rid of the convergence problems of the improper integral (1.40) and (1.41) by introducing a cut off length in order to replace the improper integrals by proper integrals over a finite range. With the value of the Planck-length of $L_P = \sqrt{\frac{h \cdot G}{c^3}} \approx 4.05 \cdot 10^{-35} m$ [Tip 03], the integral then comes to the value of (1.49).

(The factor of 2 in front of the integral represents two possible states of polarisation.)

$$\left.\frac{E}{V}\right|_{GD} = 2 \cdot \int_{|\vec{k}|=\frac{2\pi}{\infty}}^{\frac{2\pi}{L_P}} E_0(|\vec{k}|) \frac{d^3k}{(2\pi)^3} = 2 \cdot \int_{|\vec{k}|=\frac{2\pi}{\infty}}^{\frac{2\pi}{L_P}} \tfrac{1}{2}\hbar\omega \frac{d^3k}{(2\pi)^3} = \underbrace{2 \cdot \int_0^{\frac{2\pi}{L_P}} \tfrac{1}{2}\hbar c|\vec{k}| \cdot |\vec{k}|^2 \frac{dk}{2\pi^2}}_{\text{Transformation into spherical coordinates according to [KUH95]}}$$

$$= 2 \cdot \frac{\hbar c}{4\pi^2} \cdot \frac{1}{4}\left(\frac{2\pi}{L_P}\right)^4 = \frac{2\hbar c \pi^2}{L_P^4} = 3.32 \cdot 10^{+113} \frac{J}{m^3} \qquad (1.49)$$

How can the discrepancy between (1.49) and (1.44) be interpreted ?

Well – the result of (1.49) is mostly regarded very sceptically, because there is not a serious argument for the way, how the convergence problems have been suppressed. In principle the idea of our model also begins with (1.30) same as the Geometrodynamical model, but our model has a serious physical argument to solve the convergence problems.

In principle, we could interpret the result in a way, that there is no contradiction at all. The model of the present work calculates only one part of the energy-density of the space, whereas the Geometrodynamical value gives the total energy-density of the space. So our values should be smaller than the Geometrodynamical value – and this is the case. But nevertheless, it is hard to imagine, that the difference is that large. Probably the discrepancy can not be dissolved that easy.

Totally different is the interpretation of the values from cosmology and astrophysics in (1.48): The total energy density of cosmology should be larger than the value of the present work, giving only part of the energy density. But we observe the contrary. If the value of our model is larger than the value of (1.48) – can both values be possible ? The problem is gravitation: If our value would be correct – shouldn't this mean that the attractive forces of gravitation should be larger than the observed values of cosmology ?

In reality, it is not that simple to construct a contradiction. Reason: On the one hand the question about the accelerated expansion of the universe is still unsolved [Cel 07] (and there are symptoms that the expansion of the universe might be decelerated). On the other hand, expansion of the universe has to presume a special distribution of matter and dark matter – normally a distribution inside a sphere with a radius taken from the hypothesis, that the beginning of the universe was the "big bang" and the expansion of the matter and the dark matter is for sure not faster than the speed of light [Per 98], whereas the energy density of the zero point oscillations refer to the whole space $\square^3$. Hence there is also energy (and thus mass) in the space outside the sphere which we call universe. For gravitation, this difference is definitely important.

In this sense, there are too many open questions to see already a contradiction between the cosmological value of the energy density of the vacuum and the values of the model presented here. An interpretation could be like this:

Obviously, the calculation of the energy density of the zero point oscillations of the vacuum runs into convergence problem with an improper integral, leading to an infinite energy density. Classical Quantum theory goes around this problem by ignoring the infinity and fixing the origin of the energy scale (energy zero) on top this infinite energy [Lin 97]. This perception is not fully satisfying, so Geometrodynamics tries to cut off the improper integral artificially by taking a finite range for the integration. It is clear that this approach avoids problems of divergence by principle, but it results in rather strange values, which are regarded with scepticism. A physical solution for this divergence problem is found in the work presented here, and it leads to sensible values for a certain part of the energy density, namely for this part which comes from the electromagnetic zero point oscillations. But it does not calculate the total energy density of the vacuum.

4. Experiments to convert vacuum-energy into classical mechanical energy

4.1. Concept of an electrostatic rotor

Now we want to find a possible setup, with which we can extract some energy from the energy-flux of the circulation of the electrostatic field energy and the vacuum-energy (as described in sections 2 and 3). We want to convert this extracted energy into a classical type of energy, which can be experienced directly in the laboratory. If it would be possible to invent a rotor, which is driven by this energy-flux, this would be a direct experimental verification of the energy-flux.

A possible constellation, which fulfils this task, can be seen in fig.6. This is a principle sketch, which is made for sake of simplicity in order to be easy understandable, although this design does not yet have the maturity to be ready for the practical realization. For instance, it is impossible by principle to realize a punctiform field source, because the electrical charge has to be put onto a body of real existing material, but the fundamental explanations and considerations on the basis of this model can be understood very clear and easy.

Thus, we want to begin section 4.1 with a detailed analysis of the rotor drawn in fig.6: We want to calculate the forces acting on the rotor blades and the torque driving the rotor, coming from the electric field of the field source. Later, in section 4.2, we will further develop this rotor into a configuration suitable for a real experiment.

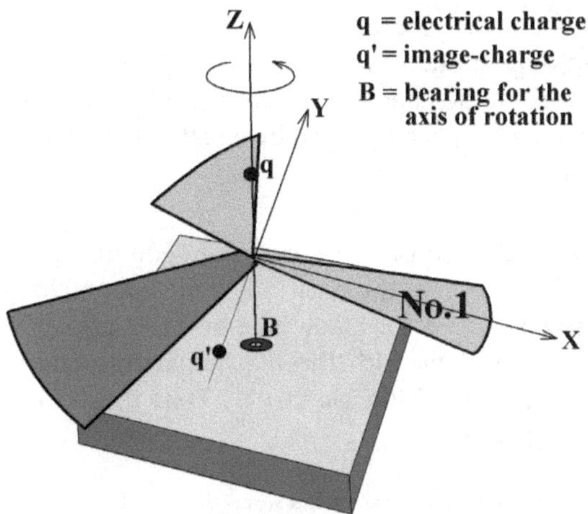

Fig. 6: Rotor consisting of three rotor blades, of which each is tilted by an angle of 45° against the xy-plane. The axis of rotation is the z-axis, which might be mounted on the bearing "B". Also on the z-axis, but above the rotor, the field source q is mounted, whose Coulomb-field acts on the rotor blades, causing Coulomb-forces, which permanently drive the rotor – as long as the practical setup is built in a way, that forces of friction will be smaller than the driving Coulomb-forces. (In section 5 we will see some further practical conditions for the occurrence of a rotation.)

In order to determine the Coulomb-forces produced by the charge q, acting onto the rotor blades, it is sufficient to determine the forces acting onto one rotor blade, because of the symmetry of the setup. The orientation of the Cartesian xyz-coordinate system is done in such a way, that the x-axis is exactly the middle-line of the rotor blade, which we take for our calculation. This rotor blade will be blade named as no.1 for further considerations.

The drive of the rotor by vacuum-energy is possible, because the charge q permanently emits field energy, which is continuously converted into mechanical energy, which has its reason in the fact, that the emitted electrical flux[2] (as illustrated in textbooks by the use of flux-lines) is stopped or redirected by the metallic surfaces of the rotor blades, and this stop or redirection causes mechanical forces acting onto the rotor blades. These driving Coulomb-forces can be calculated by the use of the image-charge method [Bec 73], [Jac 81]. Thus in fig.6 the image-charge q' corresponding to the charge q with regard to rotor blade no.1 is marked.

---

[2] The electric flux Φ_e can be defined in analogy to the magnetic flux $\Phi_m = \int_C \vec{B} \cdot d\vec{A} = \mu_0 \cdot \int_C \vec{H} \cdot d\vec{A}$ through a closed area C as $\Phi_e = \varepsilon_0 \cdot \int_C \vec{E} \cdot d\vec{A}$.

The Coulomb-force between the charge q and the rotor blade is (according to the image-charge method) the same as the Coulomb-force between the charge q and image-charge q'. Consequently, the force driving rotor blade no.1 can be simply calculated by inserting the charges q and q' into Coulomb's law.

For this purpose, we begin with a consideration of the geometry of the apparatus and with a determination of the position of the image-charge q'. Because the angle between the rotor blade no.1 and the xy-plane is 45°, this rotor blade defines a plane with the mathematical equation $z := z(x,y)$ which has the mathematical function $z = -y$. Thus the position vectors of the points of this plane are

$$\vec{r} = \begin{pmatrix} x \\ y \\ -y \end{pmatrix} \quad \text{with two free parameters } x, y \in \mathbb{R} \,. \tag{1.50}$$

Because of the symmetry of the assembly, the considerations for the determination of the forces do not alter by principle, when the rotor-blades rotate during time. Also because of the symmetry, the forces are analogously for all three rotor-blades. Thus, it is sufficient to calculate the force and the torque in the moment of consideration chosen here, and to do this calculation just for the blade no.1 and to transform the results to all three rotor blades later. In any case, the axis of rotation is the z-axis, so that all rotor-blades move within the xy-plane.

A preliminary work (before the calculations of the forces onto rotor blade no.1 can start) is the determination of the position of the image-charge q' with respect to the plane of (1.50).

The charge q is located at the z-axis at the z-coordinate z_0. The location of the image-charge q' can be found as drawn in fig.7, where we look from the direction of the x-axis onto the xz-plane. In this view, the rotor blade no.1 is drawn as a straight line with the mathematical function $z = -y$, which we already know from the parameterization of the mathematical function of the plane of the rotor blade. The position of the image-charge q' is then on the y-axis at the coordinate $y = -z_0$. The x-coordinates of q and q' remain zero, so the position vectors of the charge and the image-charge are

$$\vec{r}_q = \begin{pmatrix} 0 \\ 0 \\ z_0 \end{pmatrix} \quad (1.51)$$

for the position of the charge q

and $\vec{r}_{q'} = \begin{pmatrix} 0 \\ -z_0 \\ 0 \end{pmatrix}$ for the position of the image-charge q'. (1.52)

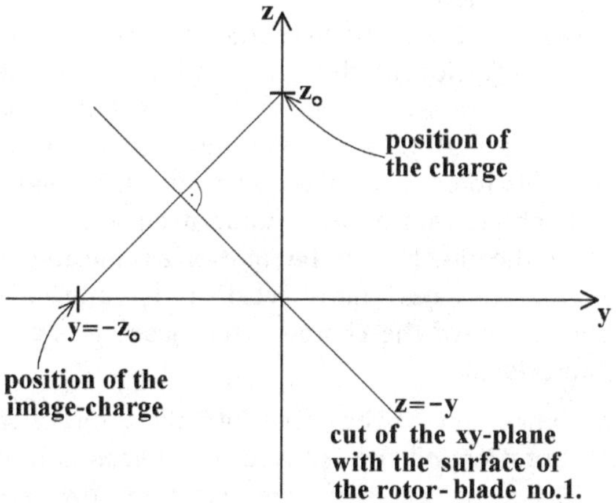

Fig. 7: Sketch for the determination of the position of the image-charge. The charge q as well as the image-charge q' have the x-coordinates $x=0$. Thus it is sufficient, to construct the position of the image-charge using a two-dimensional cut of the three dimensional geometry, namely the xz-plane.

For now we know the positions of the field-source q and its image q', we can easily calculate the forces onto the rotor-blades, just by applying Coulomb's law. Therefore, we put the value of $q'=-q$ for the image-charge into the formula. So the Coulomb-force between the charge and the image-charge (which is identically the same as the force between the charge and the rotor blade) can be written as

$$\vec{F} = \frac{1}{4\pi\varepsilon_0} \cdot \frac{(+q)\cdot(-q)}{|\vec{r}|^2} \cdot \vec{e}_r \quad (1.53)$$

with $\vec{r}$ = vector from q' to q and $\vec{e}_r$ = uint-vector in direction of $\vec{r}$,

where $|\vec{r}| = \sqrt{2} \cdot z_0 \Rightarrow |\vec{r}|^2 = 2 \cdot z_0^2$ and $\vec{e}_r = \begin{pmatrix} 0 \\ \sqrt{1/2} \\ \sqrt{1/2} \end{pmatrix}$ with $|\vec{e}_r| = 1$.

(1.54)

$$\Rightarrow \quad \vec{F} = \frac{1}{4\pi\varepsilon_0} \cdot \frac{-q^2}{2z_0^2} \cdot \begin{pmatrix} 0 \\ \sqrt{1/2} \\ \sqrt{1/2} \end{pmatrix} = \frac{-q^2}{\sqrt{128} \cdot \pi\varepsilon_0 \cdot z_0^2} \cdot \begin{pmatrix} 0 \\ 1 \\ 1 \end{pmatrix}$$ for the force between charge and image-charge.

The crucial point is: The charge q as well as the rotor-blade feels a component of the force in y-direction (the force onto the rotor blade has the opposite algebraic sign as the force onto the charge because of "actio = reaction"), which causes a rotation of the rotor-blades around the z-axis, because the force has a tangential component with regard to the movement of the rotor blade around the z-axis. For illustration, we can again have a look to fig.6. This force is attractive, because the image-charge has the opposite algebraic sign as the charge itself. From this consideration, we understand the direction of the rotation as indicated in fig.6 with a bent arrow around the z-axis. This direction of the rotation is independent of the algebraic sign of the charge q (because q and q' always hat opposite algebraic signs).

An additionally existing component of the force into the z-direction should not influence the rotation of the rotor blade, because it is absorbed by the bearing "B". In the real experiment (section 5) we will later see, that this z-component of the force together with a radial component of the force (here in the direction of x) nevertheless causes practical technical problems depending on the type of bearing used.

In order to develop a feeling for the forces expected in the practical setup, an exemplary calculation with some arbitrary but realistic dimensions shall be demonstrated, for instance as done in fig.8. There we see the topview from the z-axis onto an exemplary rotor rotating in the xy-plane. Realistic dimensions also demand a realistic size field source, which is now a sphere with a real diameter.

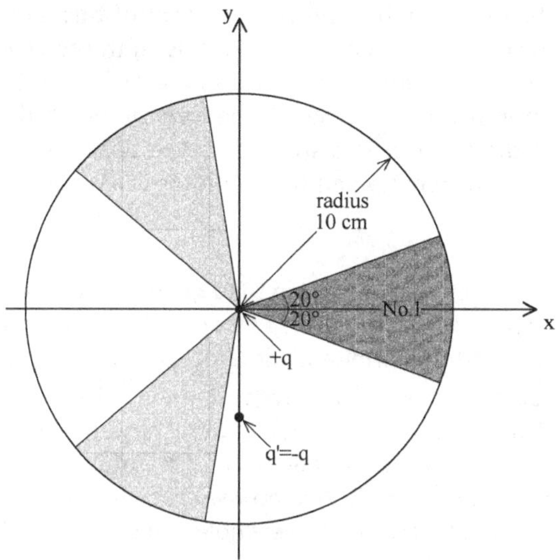

Fig. 8: Rotor with three blades and a diameter of 20 centimeters, rotating around the z-axis. In the picture, we see the projection of the rotor onto the xy-plane together with some values for the angles covered by the metal-blades. The charge q is mounted on the z-axis at the position $z_0 = 5cm$, so that the image-charge q' turns out to be on the y-axis at the position $y = -z_0 = -5cm$.

For a really existing setup, the electrical charge q has to be put onto some really existing matter. Let us chose an electrically conducting sphere with a diameter of $2R = 1.0\,cm$, and let us mount its centre at the position $z_0 = 5cm$. The capacity of such a spherical capacitor (against infinity) is $C = 4\pi\varepsilon_0 \cdot R$. If we put this sphere to an electrical voltage of $U = 10kV$ (which should be a good value in order to avoid electrical breakthrough), it will take an electrical charge of $q = C \cdot U = 4\pi\varepsilon_0 \cdot \underbrace{R}_{0.5cm} \cdot \underbrace{U}_{10kV} \approx 5.56 \cdot 10^{-9} C$. The image-charge thus will have a value of $q' \approx -5.56 \cdot 10^{-9} C$.

Putting these values into (1.54) for the force between the charge and the image-charge, we come to

$$\Rightarrow \vec{F} = \frac{-q^2}{\sqrt{128} \cdot \pi\varepsilon_0 z_0^2} \cdot \begin{pmatrix} 0 \\ 1 \\ 1 \end{pmatrix} = 3.93 \cdot 10^{-5} N \cdot \begin{pmatrix} 0 \\ 1 \\ 1 \end{pmatrix} \qquad (1.55)$$

The z-component of this force, which is parallel to the direction of the axis of rotation will not be recognized (and not be important at all, besides for technical problems later), but the y-component of this force directly causes the rotation of blade no.1 around the z-axis (if this force is strong enough to overcome the force of friction), because it acts tangentially onto the rotor blade. Because of the symmetry of the assembly, the forces onto the other blades are understood analogously. This means that the principle of functioning of the motor is already explained now.

It should be mentioned, that the force according to (1.55) is only a rough approximation, because our calculation of $\vec{F}$ up to now gives the total force between the charge q and the infinite plane $z := z(x,y) = -y$. In reality, our rotor is not infinite but it is finite, so we have refine our calculation. For the determination of the force actually working on the rotor-blade covering a finite part of this plane, we again want to turn our attention to fig.8 showing a projection of the assembly. And now we have to calculate which percentage of the electric flux through the whole plane $z := z(x,y) = -y$ will pass the finite blade from which we can determine the force onto this finite blade. Let us begin this calculation by writing the electrostatic potential of the charge and the image-charge, valid for the space between the plane $z := z(x,y) = -y$ and the charge q:

Coulomb's potential of the charge q is $V = \frac{1}{4\pi\varepsilon_0} \cdot \frac{q}{d}$,

and Coulomb's potential of the image-charge q' is $V' = \frac{1}{4\pi\varepsilon_0} \cdot \frac{q'}{d'}$,

with $d = \sqrt{(\vec{r} - \vec{r}_q)^2} = \sqrt{x^2 + y^2 + (z - z_0)^2}$ and $d' = \sqrt{(\vec{r} - \vec{r}_{q'})^2} = \sqrt{x^2 + (y + z_0)^2 + z^2}$ being the distances between the charge respectively the image-charge and the space point, at which the potential has to be calculated. From there (and because of $q' = -q$), we come to the total potential within the space between the charge and the plane $z := z(x,y) = -y$, and we find:

$$V_{tot} = V + V' = \frac{1}{4\pi\varepsilon_0} \cdot \frac{q}{\sqrt{x^2 + y^2 + (z - z_0)^2}} + \frac{1}{4\pi\varepsilon_0} \cdot \frac{-q}{\sqrt{x^2 + (y + z_0)^2 + z^2}} . \qquad (1.56)$$

The electrostatic field strength is calculated as usual: $\vec{E} = -\vec{\nabla} \cdot V_{tot}$. (1.57)

From there we calculate the percentage of the electric flux through the rotor-blade relatively to the electric flux through the total plane $z := z(x,y) = -y$, where the last mentioned is of course a convergent improper integral. This was done in a numerical approximation, leading to a result of about $(4\pm0.5)\%$. This means, that the force acting onto the finite rotor-blade can be estimated to be about $(4\pm0.5)\%$ of the total force $\vec{F}$, with a precision sufficient for the first planning of an experiment. Consequently, we get the y-component of the force acting on each single finite rotor-blade as

onto each single rotor blade $F_y \approx 3.93\cdot10^{-5}N\cdot4\% \approx 1.6\cdot10^{-6}N$, (1.58)

and thus the force onto all three rotor blades $3\cdot F_y \approx 4.7\cdot10^{-6}N$, (1.59)

as far as causing a torque driving the rotor.

At least we want to know the torque, with which the charge q turns our rotor-blades. Therefore, we have to take into account, that the force $3\cdot F_y$ does not act onto one single point, but its action is distributed along all points of which the surface of the rotor blades consist. The calculation of the torque is a simple mechanical problem, which does not need a detailed demonstration here. Its result is a torque of about

$$\left|\vec{M}_{tot}\right| \approx 9\cdot10^{-8}\,Nm\,, \qquad (1.60)$$

which is given as "approximately" because of the numerical approximation of (1.58) and (1.59). This example of calculation demonstrates distinctly, to use a bearing with very low friction. The realization of such a bearing will be an important task for the practical realization in the experiment.

The physical principle of the electrostatic driven rotor, which converts vacuum-energy (from the flux of the propagation of an electrostatic field) into mechanical energy of a rotation, is now invented. **It shall be emphasized here that the development of the functioning principle only needs two fundamental assumptions: One is the validity of Coulomb's law and the other one is the validity of the image-charge method.**

As soon as the electrical charge is mounted above the rotor, the rotor is accelerated until the forces of friction (or the forces of a mechanical utilization) are the same strong as the driving Coulomb-forces. When this condition is reached, the rotor spins with constant revolution speed (angular velocity).

4.2. First experiments for the conversion of vacuum-energy

With regard to the expectation of a very small torque, the geometry of the practical setup for an experiment has to be optimized, with two aspects being of special importance, namely a maximization of the driving torque and a minimization of the friction. Otherwise, the driving torque would not surmount friction at all, resulting in a standstill of the rotor.

A very simple rough estimation already allows to recognize, that a rotor with a diameter of $20cm$ and a torque as given in (1.60) of less than $10^{-7} Nm$ is not capable to surmount the friction of a normal standard ball bearing. So we have to regard the two optimizations as said above:

<u>**First part:**</u> Maximization of the driving torque:

The spatial distribution of the field producing charge, i.e. the shape and the position of the field source, has not been optimized up to now, so that there is a lot of possibility for optimization. For the purpose of the calculation of the torque produced by different types of field sources, two different types Finite-Element-Algorithms have been used. The first algorithm was self-made [Pas 99]; it calculates the forces on the basis of finite charge elements (on the surface of the field source) and finite image-charge elemets on the rotor blade using Coulomb's law giving the forces between those pairs of charges. The second algorithm was the commercial FEM-program ANSYS [Ans 08], whose working is based on potential theory. Two applications of two different algorithms - this has the purpose to provide a possibility the check the results.

The self-developed algorithm is based on the application of the image-charge method and on Coulomb's law according to section 4.1. Therefore, the field source was subdivided into finite charge elements and the rotor blade was subdivided into finite image-charge elements, so that finite Coulomb-forces between all pairs of charges and image-charges have been calculated. For each finite Coulomb-force, the torque has been calculated taking its radius of rotation into account. The sum of all these finite torques gives the total torque driving the rotor. A check on self-consistency was done by calculating the electrical potential and the electrical field-strength according to (1.57), which additionally provides the possibility to assure, that there will not be a field strength above the electrical breakthrough in the practical experiment.

For the Finite-Element-Program ANSYS, the input does not consist of a rotor and a field source, but it consists of the space filled with field. Those positions which are occupied by the rotor and the field source are given as boundary conditions, namely as an electrical potential or as an electrical charge located at the surface of the space being modelled. ANSYS calculates the potential and the field strength in space being filled with finite elements, and from there force- and torque- values on the surfaces of the model can be calculated.

Because the algorithm ANSYS had limited availability [Ihl 08], the optimization was done with the self-developed algorithm. Therefore, the geometry of the field source and of the rotor have been changed as well as the geometrical location of both, relatively to each other. ANSYS was only used to check the results of the own algorithm. Both methods are numerical approximations, both converge with increasing number of finite elements, and both come to the same results within their numerical uncertainty. So ANSYS confirmed the results of the self-made algorithm.

It was found that a flat disc is a good field source if its diameter is a bit larger than the diameter of the rotor. An example therefore is shown in fig.9, which produces a torque of approximately $M=1.2\cdot 10^{-5}\,Nm$, which is remarkably larger than the torque of the simple setup in section 4.1, fig.8. And the electrical voltage to reach this torque in fig.9 is only $U=7kV$ between field source and rotor. A field source, which is a little bit larger than the rotor, is sufficient with regard to the torque. A field source much large than the rotor does not further enhance the torque.

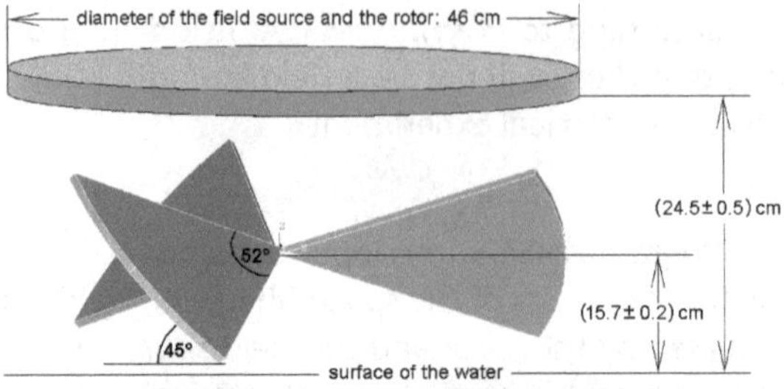

A torque of about $M \approx 1.2 \cdot 10^{-5}$Nm is being calculated.

Fig. 9: Principle design of an electrostatic rotor for the conversion of vacuum-energy into mechanical energy. The disc (red) is electrically charged (field source), and the rotor (blue) is connected to ground.
The specified dimensions represent the values in the setup of a really conducted experiment

Furthermore, it was found that a hole in the centre of the field source, as shown in fig.10 does not remarkably reduce the torque, so that it would not be a problem to lead an axis through the centre of the field source. This observation is plausible, because the dominant part of the torque is of course produced by those elements of the surface, which have large radii of rotation. The option to have a hole in centre of the field source might be helpful for arrangement of the bearing.

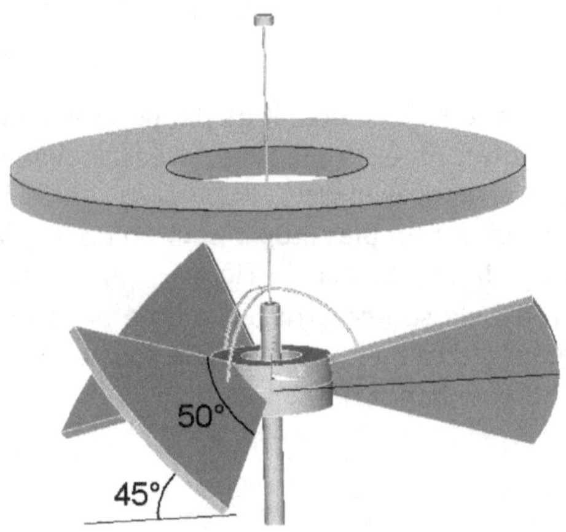

Fig. 10: This design produces about the same torque as the design shown in fig.9.

For the purpose of a further optimization of the geometry, a systematic variation of the parameters "voltage" and "rotor radius" have been analyzed, leading to the following proportionalities:

- Torque $M \propto U^2$, and thus:

  driving power $P \propto U^2$ (with U = electrical voltage)
- Torque $M \propto R^2$, and thus:

  driving power $P \propto R^2$ (with R = diameter of the rotor).

These proportionalities have been deduced while adjusting the distance between rotor and field source in such manner, that the maxima of the field strength have been kept below a constant limit (below the electrical breakthrough of air). This has the sense to avoid electrical breakthrough.

With the knowledge of these proportionalities, it is possible to get a quick estimation of the size of the rotor necessary for a requested torque.

Second part: Minimization of the friction of the bearing:

For the first approach, it is quite usual to think about a mechanical bearing. But it is necessary to be very careful to minimize its friction, and it causes several further problems, which will be topic of section 5.

For a mechanical tribological pairing, the forces of friction F_R are proportional to the force F_N normal (perpendicular) to F_R (in the plane of the surfaces in contact). In our case, this is the force with which the rotor lies on the bearing because of its mass [Stö 07]. The factor of proportionality is the coefficient of friction μ, which is μ_H for static friction at the beginning of the rotation and μ_G for dynamic friction during the rotation. So the force of friction is

$$F_R = \mu \cdot F_N \ . \tag{1.61}$$

The normal (perpendicular) force F_N, is given simply by

$$F_N = m \cdot g \ , \tag{1.62}$$

with m = mass of the rotor and g = gravitational acceleration.

The torque with which the mechanical bearing brakes the rotation is then

$$M_R = r \cdot F_R \tag{1.63}$$

with r = radius with which the forces of friction act at the bearing.

Putting (1.61) and (1.62) into (1.63) leads to the expression of the torque

$$M_R = r \cdot F_R = r \cdot \mu \cdot F_N = r \cdot \mu \cdot m \cdot g \ . \tag{1.64}$$

This is the basic formula to analyze and compare the operational capability of several types of bearings with regard to the experiment of the electrostatic rotor to convert vacuum-energy into classical energy of rotation. The goal of the following comparison will be the minimization of the friction. Only if it is possible to reduce static friction to a torque below the value of fig.9, it is sensible to conduct the experiment.

A reasonable comparison requires the comparable preconditions. Let us use the input of $m = 8.7\,Gramm = 8.7 \cdot 10^{-3}\,kg$ (this is the value in the experiment described soon) and $g \approx 10\,{}^{m}\!/_{s^2}$ (sufficient for the present estimation). The values of r and μ are characteristic for each type of bearing, thus they are to be varied for the purpose of comparison.

(a) Ball bearing:

A typical value for the rolling friction of steel spheres within a ball bearing is for instance $\mu_R = 0.002$ [Dub 90]. If a ball bearing has a radius of $r = 1\,cm$ the braking torque would be according to (1.64):

$$M_R \approx \underbrace{10^{-2}\,m}_{r} \cdot \underbrace{2 \cdot 10^{-3}}_{\mu} \cdot \underbrace{8.7 \cdot 10^{-3}\,kg}_{m} \cdot \underbrace{10\tfrac{m}{s^2}}_{g} \approx 1.7 \cdot 10^{-6}\,Nm\,. \tag{1.65}$$

(b) Toe bearing:

Here the rotor is running directly on the tip of the toe bearing, so that the coefficient of friction is not for rolling friction but for static friction, for which a typical value between two steel surfaces can be for instance $\mu_R = 0.3$ [Dub 90]. The main advantage of a toe bearing is the very very small radius with which the force is transmitted, so that the torque is very small. From the observation of toes with a microscope, typical values of the toe radii in the order of magnitude of about $r = 10\,\mu m$ have been found. If the radii are even smaller (which can be manufactured), the toe might not be strong enough to withstand the weight of the rotor. This leads to an estimation of the braking torque in the case of a toe bearing in the order of magnitude of

$$M_R \approx \underbrace{10^{-5}\,m}_{r} \cdot \underbrace{0.3}_{\mu} \cdot \underbrace{8.7 \cdot 10^{-3}\,kg}_{m} \cdot \underbrace{10\tfrac{m}{s^2}}_{g} \approx 2.6 \cdot 10^{-7}\,Nm\,. \tag{1.66}$$

If the rotor has less weight, the braking torque is even smaller.

(c) Bearing by swimming on the surface of a fluid (hydrostatic bearing):

There are many fluids (liquids and gases) used as lubricants with low friction in order to reduce tribological forces. Gases are known to produce extremely low friction as for instance in the application of technical air suspensions (for example air conveyor tables), but in the present experiment they are not applicable, because they produce a gas flow which might drive the rotor. In order to exclude such artefacts, the bearing on a gas is not used. Liquids can also allow rather low friction. Especially water is known to be a liquid whose friction converges to zero asymptotically as soon as the speed of the relative movement goes to zero (see fig.11, following [Hei 97]). In principle, we know this from everyday's life: A ship of several hundred or thousand tons can be moved it we push it with a hand, as long as the speed of movement is very slow. This leads us to the following situation:

$$M_R \approx \underbrace{20cm}_{r} \cdot \underbrace{(\mu \to 0)}_{\lim_{v \to 0}} \cdot \underbrace{8.7 \cdot 10^{-3} kg}_{m} \cdot \underbrace{10 \tfrac{m}{s^2}}_{g} \approx \text{very small for slow movement} \quad (1.67)$$

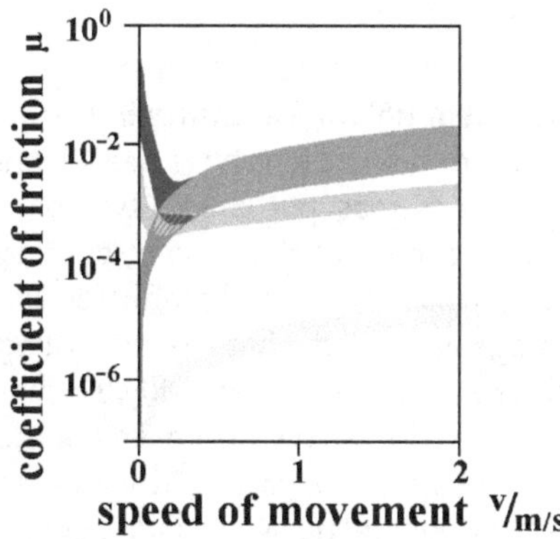

Fig. 11: *Dependency of the coefficient of friction μ of the speed of movement for different types of bearings:*
- blue and purple → hydrodynamic bearing
- red and purple → hydrostatic bearing
- green → rolling bearing
- yellow → aerostatic bearing
For the hydrostatic and the aerostatic bearing, the coefficients of friction converge to zero if the speed of movement is very slow.

Consequently, the comparison of the low friction bearing comes to the resumée:

According to (1.65), (1.66) and (1.67) all three analyzed types of bearing should applicable for the planned experiment in principle, because they produce a braking torque less than the driving torque of $M=1.2 \cdot 10^{-5} Nm$ (see fig.9). But as a matter of fact, the input values for the comparative analysis are ideal conditions. It is not possible to predict, how real condition deviate from such ideal conditions. This was the reason for the following approach: Let us start with the lowest friction, the fluid bearing (= hydrostatic bearing). After this will be successful done, it is possible to expand the experiment to other bearings.

Hydrostatic bearing does not allow a large speed of rotation, but if there is a small driving torque at all, it allows low speed of rotation. And this is enough for the verification of the principle of the conversion of vacuum-energy (section 4). After the successful completion of the conversion of vacuum-energy with the hydrostatic bearing, it is possible to start with a toe bearing, because the hydrostatic bearing is not the first choice for technical applications. But the development of a rotor with a toe bearing to technical maturity is still under optimization (section 5), and threre are several questions to be answered.

The first realization of the hydrostatic bearing with a swimming rotor was done on the surface of water (later other realizations have been done on oil), at which a rotor with three blades was put onto Styrofoam floating bodies as shown in fig.12. It was found that this type of bearing has some further practical advantages, which are very helpful for the success of the experiment. One advantage is the fact, that the swimming rotor on the water surface is adjusted always exactly horizontally, so that the rotor can not tilt relatively to the field source. With the large Coulomb-forces perpendicular to the plane of rotation, this is an important advantage. But the main advantage of the swimming rotor is, that the rotor can freely move laterally on the water surface. This is important, because the attractive Coulomb-forces between the rotor and the field source initiate a self-adjustment mechanism of the rotor relatively to the field source. The attraction of the Coulomb-forces has the consequence that the rotor always finds its way into the minimum of the electrostatic potential (created by the field source). This self-adjustment mechanism is very important, because a rotor with rigidly fixed axis of rotation and imprecise adjustment (of this axis) within the electrostatic potential will only rotate until a rotor blade finds the minimum of the electrostatic potential, and at this position the rotation will stop at all. As a matter of logic, it is clear, that this condition is reached after less than one turn (of rotation) if the rotor is not perfectly adjusted. Only a rotor with very exact adjustment can follow an endless rotation (which would require a very exact manufacturing of the rotor). And the described self-adjustment mechanism effectuates always a perfect adjustment, and does this by alone (even if the assembly of the rotor is not extremely exact).

It could be mentioned that the voltage necessary to initiate the self-adjustment mechanism is less than the voltage necessary to drive the rotor, so that it is possible to enhance the voltage slowly beginning from zero, and to adjust the rotor first before the rotation begins. But in practice, there is no problem to apply a voltage from the very beginning with starts the self-adjustment and the rotation in the same moment.

Fig. 12: Photo of the electrostatic rotor under the field source as used in an experiment. It was a rather simple handmade setup, but it was capable to perform the rotation successfully and reproducible.

The practical execution of the experiment is the following:

In order to have an extremely lightweight construction (also for the minimization of friction), the rotor blades have been made of aluminium foil with a thickness of approx. $10\mu m$, which has been mounted on a frame of balsa wood with the use of cyanoacrylate adhesive glue. The rotor blades have been fixed to each other with two-component adhesive (Stabilit Express). In the centre of the rotor, an iron rod (diameter 2.2 mm) has been fixed to mark the axis of rotation. All rotor blades have been connect electrically with each other and with the iron axis with thin copper filaments (thickness approx. $60\mu m$), and the iron axis was long enough to be in electrical connection with the water. By this means the rotor was electrically grounded during the time of its rotation and the grounding did not give noticeable mechanical forces onto the rotor.

In this configuration, the rotor was put onto the surface of the water, and then the field source was mounted above the rotor with horizontal adjustment (parallel to the surface of the water). Then the water was electrically connected to ground and the field source to the high voltage supply. It is important to keep all electrical cables, which have voltage very far away from the rotor (several meters), otherwise they would cause an inhomogeneity of the electric field, which prevents the rotor from rotating. When switching on the voltage, at first the self-adjustment began and then the rotation. Fig.13 shows the observation of an exemplary measurement, at which the time was recorded (under optical observation) which the rotor needed to pass steps of rotation of 60°, i.e. its angle of rotation was $n \cdot 60^\circ \; (n \in \mathbb{N})$.

The statistical distribution of the data is due to lateral movements of the rotor during its rotation in connection with the self-adjustment mechanism of the lateral position on the surface of the water. (Of course, the convection of gas (of the air) was carefully avoided in order to exclude driving forces due to any wind.)

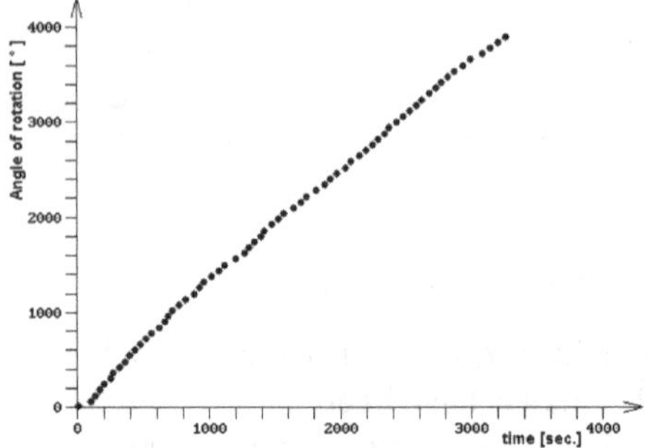

Fig. 13: Example for the measurement of the rotation of the rotor shown in fig.12 under the circumstances explained above. The angle of the rotation is given in degrees, so that one complete revolution corresponds to 360°.

Furthermore it should be mentioned, that the electrical voltage between the field source and the rotor decreased during time. In the example of fig.13, the following happened: At the beginning a voltage of $U = 7kV$ was applied, and the rotor began to rotate. After approximately half an hour, 6...8 revolutions have been fulfilled, and the data acquisition started at a moment at which the voltage was $U = 6kV$. During the following hour of data acquisition, the voltage further decreased to $U = 4.5kV$. Consequently, the angular velocity of the rotation also decreased during time, as can be seen in fig.13.

We now see a numerical estimation of the results, especially of the machine power corresponding to the energy converted from vacuum-energy:

The rotor itself has a weight of $m=8.7\,Gramm$, but three Styrofoam cuboids (each of $m=0.56\,Gramms$) additionally perform the rotation. Thus, the moment of inertia is approximately $J \approx 3.2 \cdot 10^{-4} kg \cdot m^2$. A torque of $M \approx 1.2 \cdot 10^{-5} Nm$ leads to an angular acceleration of about $\alpha \approx 2.1°/sec.^2$. The average angular velocity observed in the measurement was about $\omega \approx 0.84°/sec.$ This means that the rotor already reaches its final angular velocity after less than 0.4 seconds of acceleration. Such a short phase of acceleration could not be measured in this experiment. But the average engine power taken from the vacuum could be found to be about $P = 1.75 \cdot 10^{-7} Watt$, which finally was absorbed by the water via friction.

This was the very first proof for the operational capability of the electrostatic rotor to convert vacuum-energy into classical mechanical energy. On the one hand this is important for the fundamental science of Physics, but on the other hand it might offer a possibility for the energy supply of the future, because a precise mechanical fabrication of the setup and good electrical isolation (in order to minimize the loss of electrical charges from the field source) arises the hope, that more energy can be extracted from the vacuum.

A second experimental verification of the conversion of vacuum-energy was begun with rotors of several diameters on toe bearings [e12]. One of them is shown in fig.14. The toe bearing consists in a glass dome rotating on the tip of a steel needle. Such a bearing can be bought easily within a standard Crooke's radiometer at a glass manufacturer. The rotor blades are made of the same material as those in fig.12, of aluminium foil on a frame of balsa wood. Each of the blades has a surface of 3.5 cm x 6.0 cm; they are rotating in a plane with a distance to the field source of approx. 3.8 ... 4.0 cm; this distance is altered during the rotation).

Fig. 14: Electrostatic rotor with four rotor blades on a toe bearing, arranged below a field source made of Aluminium.

Tests with grounded rotor-blades and an electrically charged field source display the following results:

- Field source brought to a potential of 1100 Volt → 4 revolutions per minute.
- Field source brought to a potential of 1400 Volt → 12 revolutions per minute.
- Further enhancement of the voltage enhances the speed of rotation remarkably.

An exact adjustment of the field source relatively to the rotor is of great importance and has to be done very carefully, because the rigidly fixed axis of rotation does not allow the self-adjustment mechanism, which we know from the swimming rotor. Therefore the plane of the rotation has to be adjusted parallel to the field source (which is a flat disc), and the axis of rotation has to be adjusted as close as possible to the minimum of the electrostatic potential below the field source. If these conditions are not fulfilled good enough, the rotor rotates only for an angle less than 360° and then comes to a standstill at the position where it finds the minimum of the electrostatic potential. Only if the driving force (coming out of the vacuum-energy) is larger than Coulomb-force stopping the rotor in the minimum of the potential, the rotor will rotate enduring. (An angular momentum might help to overcome to minimum of the potential if the average of the driving Coulomb-force is larger than Coulomb-force stopping the rotor in the minimum of the potential.) The angular speed mentioned after fig. 14 can only be reached with very good adjustment and changes very strongly if the adjustment is changed only a little bit.

The value of the high voltage was rather moderate (the rotation begins at 1100 Volt depending on the quality of adjustment) in order to avoid the ionization of gas molecules of the surrounding air, because the recoil of gas ions had to be suppreseed as much as possible [Dem 06]. A technical application of the recoil of gas ions is known from [Bro 28] and [Bro 65] (Biefeld and Brown) and this is not what shall be observed in the present work. To be really sure to exclude the influence of gas ions, that rotor is brought into the vacuum later – and it rotates there. If should be annotated, that Biefeld and Brown use much higher field strehgth than it was done in the experiment presented here.

Very first preliminary thoughts for the exclusion of recoils of gas ions have been already presented in [e9]. Further tests are published in [e13]. The last mentioned publication shall be referenced here, its central idea is the alteration of the shape of the rotor blades as can be seen in fig.14.

- If the attractive Coulomb-forces are dominating, the rotor of fig.14 spins clockwise, same as the altered rotor of fig.15.

- But on the contrary if the forces from recoils of gas ions would dominate, the rotation would be different. Gas molecules are ionized at the regions with large field strength, which are found at the top edges of the rotor blades. Ions generated there would follow the gradient of the electrical field and be accelerated into the direction of this gradient. Thus the recoils would make the rotor of fig.14 spin clockwise but the rotor of fig.15 counter-clockwise.

- The conducted observation confirms the first mentioned behaviour going back to the dominance of the attractive Coulomb-forces, not to the recoils. This indicates that the Coulomb-forces really exist.

A definite exclusion of the recoil forces of gas ions is reported in the following section 4.3, where the gas molecules of the air have been removed by making a vacuum.

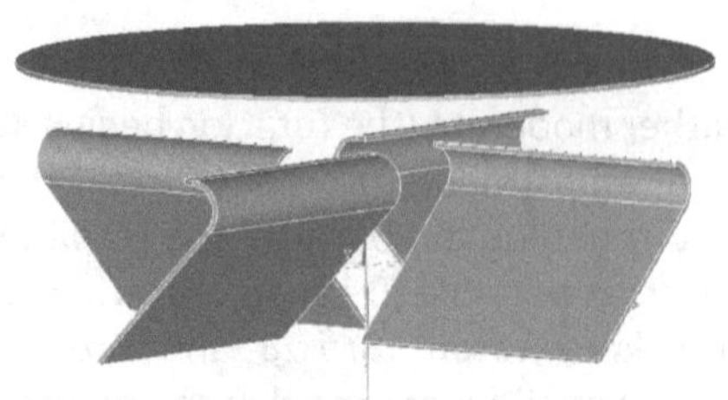

Fig. 15: The rotor of fig.14 was changed in order to analyze the influence of electrohydrodynamically induced recoils of gas ions.

4.3. Experimental verification under the absence of gas-molecules

In principle both of the following forces are under discussion in our experiment with the rotor for the conversion of vacuum-energy:

(a.) attractive Coulomb-forces (going back to the conversion of vacuum-energy),

(b.) forces due to recoils of ionized gas molecules.

The test of fig.15 does not for sure exclude the force of (a.). But the fact, that an alteration of the shape of the rotor blades (from fig.14 to fig.15) does not influence the direction of the rotation, demonstrates that the attractive Coulomb-forces are really present. The geometrical precision of the setup is not good enough to allow a quantitative estimation of both mentioned forces relatively to each other. For this purpose, the experiment had been brought into the vacuum. The quantitative estimation deduced from the vacuum experiment is reported at the end of section 4.3, and first of all: If the rotor rotates within the vacuum, the existence of the force (a.) is proven for sure. This is the purpose of section 4.3.

The reported tests under air (with room pressure) have not been regarded as a complete proof for the existence of the attractive Coulomb-forces of (a.) due to the conversion of vacuum-energy. Some colleagues did not accept this test. In order to get a proof, which is regarded as a safe verification for the attractive Coulomb-forces by everybody, the experiment of [Kna 08/09], [e14] with the rotor in the vacuum was performed. But before this experiment is described, we want to have a short glance to some other preliminary work, from which the background of the successful experiment (described afterwards) can be understood.

This preliminary work was done with a rotor made of an iron sheet rotating on toe bearing as can be seen in fig.16. The rotor has been manufactured by bending and cutting the sheet material, followed by deburring and polishing. The toe of the toe bearing was a steel needle. The setup is uncomplicated, but it is suitable for the use in a vacuum chamber.

The rotor was rotating under air (see photo of fig.16), but there is a certain risk that it falls down when the voltage is applied, because of the forces, which have to be absorbed by the bearing. Especially if the lateral adjustment of the rotor axis in the minimum of the electric potential was not exact enough, this could happen rather easy. And this is not very easy to avoid, because there is no self-adjustment mechanism as it was known from the swimming rotor. Especially inside the vacuum chamber this was a problem: As soon as the voltage is switched on, the rotor tilts and falls down.

Under air it was possible to make the rotor spin with a voltage even in the range of $2 \ldots 3kV$, coming to an angular velocity of several rounds per second. A moderate enhancement of the voltage to $3 \ldots 4 \ldots 5kV$ was enough to enhance the angular velocity up the several ten rounds per second.

Especially inside a metallic chamber (like a vacuum chamber), the absence of the self-adjustment mechanism is a serious problem, because the electrically conducting walls of the chamber influence the electric field drastically, so that an adjustment of the rotor in the minimum of the electrostatic potential is hardly possible. In this context we shall remember, that even an electric cable influences the electric field so much, that the rotation will be disturbed. Thus, the setup in fig.16 will lead to a rotation of few angular degrees, followed by a fall down of the rotor. This is the observation inside the vacuum chamber filled with air in the same way as inside the evacuated vacuum chamber. The rotor of fig.16 falls down, but the rotor of fig.14 tilts inside a metallic chamber. The rotor of fig.14 does not fall down inside the chamber because the glass dome prevents the fall. If the rotor tilts too much, it can not rotate because the force of friction between bottom end of the glass dome and the steel needle is to strong. This demonstrates that the electrostatic rotor reacts very sensitive to the presence of the walls of a metallic chamber.

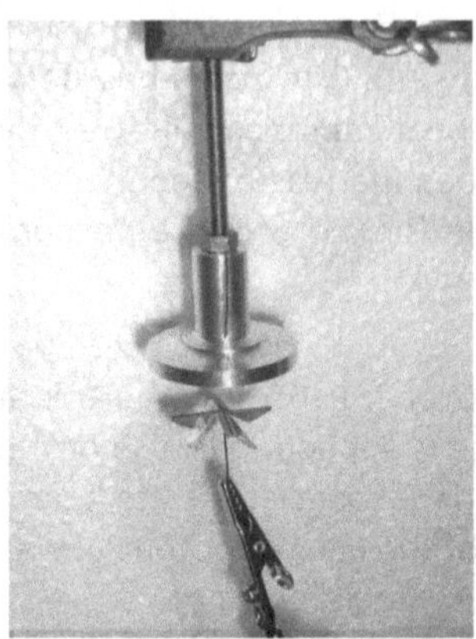

Fig. 16: Electrostatic rotor made from iron sheet material. The rotor has a diameter of about 2.5 *cm* *. The field source is made of Aluminum. It is connected with the high voltage from the very top, because it is not allowed that the rotor sees any electrical cables (they disturb the electrical field). The grounding of the rotor is done from the bottom also with cables far away from the rotor. The toe bearing allows an inclination of the plane of rotation, which can be large enough that the rotor falls down. This problem is not yet solved, so that the rotor really used in vacuum experiment was built on a different type of bearing.*

After these tests it became clear, that it is the most effective method for the verification of the conversion of vacuum-energy, to use the self-adjustment mechanism even for the rotor in the vacuum chamber. To check this statement, a small rotor for an experiment inside a vacuum chamber was manufactured (see fig.17) and preliminary tested swimming on water in a simple metallic chamber without the possibility to evacuate. This was the first rotor spinning inside a metallic chamber. The self-adjustment mechanism began to work at about 1 kV and the rotation began unproblematic at about 2 kV with approx. 1 turn in 3...5 seconds. The angular velocity could be enhanced up to several turns per second at a voltage of about 4 kV.

Fig. 17: Rotor made of Aluminium foil on a frame of balsa wood, swimming on four floating bodies of Styrofoam on the surface of water inside a metallic chamber. With the self- adjustment mechanism working, the rotation was no problem inside the metallic chamber.

In order to transfer this successful principle into the vacuum, the water had to be replaced by a fluid with low vapour pressure. Therefore a special vacuum-oil was used with the name „Ilmvac, LABOVAC-12S", which has a vapour pressure of $10^{-8} mbar$ [Ilm 08]. Not perfectly ideal is the rather large viscosity of this oil (dynamic viscosity at 40°C of $94 milli Poise$ according to manufacturer information), which is more than two orders of magnitude larger than the viscosity of water (dynamic viscosity at 40°C of $0.65 milli Poise$). Thus, the swimming rotor on the oil needs remarkably larger force and torque than the swimming rotor on water if it shall rotate. The consequence is that the voltage to drive the rotor on oil has to be larger than the voltage applied to the rotor on water, and furthermore the rotor on oil goes much slower than the rotor on water. At pre-tests in air, two rotors, one with a diameter of $51 mm$ and the other one with a diameter of $58 mm$ have been tested. The voltage necessary to achieve rotation with the rotor on oil was minimally about $8 \ldots 12 kV$ in comparison to a voltage of about $1.5 \ldots 2 kV$ at which the rotation of the rotor on water begins.

And the angular velocity of the rotor on oil was only about one turn per 2 ... 3 hours in comparison with an angular velocity of several turns per second of the rotor on water. The angular velocity of about 2 ... 3 hours per one turn will be brought back to our mind later when the rotor is investigated in the vacuum.

The vacuum oil did not have a measurable conductivity (with our Ohmmeter), so the electric grounding of the rotor blades has been done with a thin copper filament (diameter $60 \mu m$) from the rotor to the bottom of the metallic chamber, where is was not rigidly fixed, but it was allowed to slide. The Styrofoam floating bodies (see fig.17) have been replaced by balsa wood floating bodies, which have been sealed with blue lacquer in order to avoid permeation of the oil into the balsa wood. Because of the viscosity of the oil, the copper filament slows down the angular velocity of the rotor a little bit, but it did not prevent the rotation, so it was accepted. The rotor with balsa wood floating bodies can be seen in fig.18.

Two types of driving forces might possibly occur, namely (a.) attractive Coulomb-forces in connection with the conversion of vacuum-energy and (b.) recoil forces of ionized gas molecules (as said above). The force of (a.) is possible in air as well as in the vacuum, whereas the force (b.) is only possible at the tests with air. If the force (b.) would not exist at all, the rotor should have the same angular velocity in air as in the vacuum (if the electric field is the same). If the force (b.) exists additionally to (a.), the angular velocity in air should be larger than in the vacuum (with the same electric field in both cases). If the force (a.) would not exist at all, the rotor should rotate only in air but not in vacuum.

Fig. 18: Photo of a $58mm$ - rotor with four rotor blades of aluminium-foil fixed on balsa-wood with adhesive (same as in fig.17). The balsa floating bodies have been covered with blue lacquer in order to avoid permeation of the oil into the balsa wood. The aluminium rotor blades are connected with a copper wire, which is fixed at the axis of rotation in the centre of the rotor and from there it goes down to the ground plate of the metallic chamber so that the rotor blades can be grounded.
After the successful pre-tests inside a metallic chamber under air, two rotors (with diameters of $51mm$ (fig.19) and with $58mm$ (fig.18)) have been brought into a vacuum chamber with a diameter of $100mm$. The field source has a diameter of $63mm$, is made of Aluminium and can be seen in fig.20.

Fig. 19: Photo of an electrostatic rotor in the unclosed vacuum chamber, swimming on vacuum oil. The floating body sealed with blue lacquer consists of one piece of balsa wood in this example, so that the copper wire for the electrical grounding is conducted through the middle of the floating body down to the ground of the vacuum chamber.

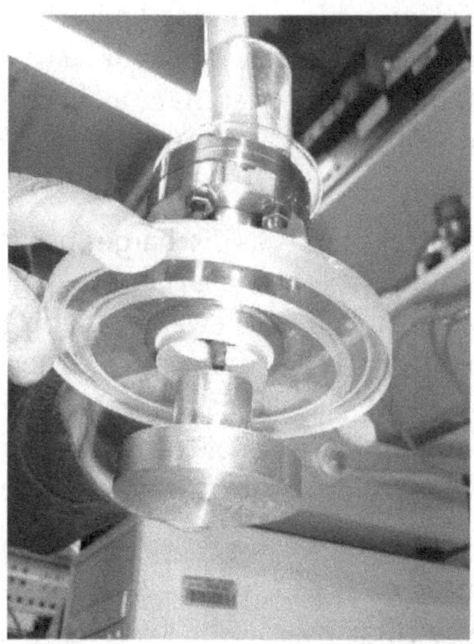

Fig. 20: The field source produced of Aluminium is mounted at the top flange of the vacuum chamber from where it is supplied with a high-voltage duct. The voltage applied to the field source can be arranged from $0 \ldots 30kV$ *.*

The procedure of the experiment, at which finally the rotor rotated inside the vacuum, was the following:

After the rotor was inserted into the vacuum chamber, the top flange with the field source was closed, bringing the field source into position. Then high voltage was applied (in the range of $10 \ldots 20kV$) to test the rotation in air. At this phase of the experiment two forces driving the rotor were imaginable, the attractive Coulomb-force as well as force from the recoils of ionized gas particles. The electrostatic Coulomb-force does not produce an electric current, but forces from the recoils of ions do. At the begin of the pumping process an electric current could be measured, especially when the pressure passed the range at which according to Paschen's law ionization is to be expected rather much [Ker 03], [Umr 97]. This is especially the case for a gas pressure between few $10\,mbar$ until down to few tenths of a $mbar$, which is well in agreement with Paschen's law: At a distance of about $19 \ldots 20\,mm$ between the upper edges of the rotor blades and the field source, the Paschen-minimum of the breakthrough voltage (with $p \cdot d = 7.5 \cdot 10^{-6}\,m \cdot atm$, where $p =$ pressure and $d =$ distance of the capacitor plates) occurs at a pressure of about $p \approx 0.4\,mbar$. This means, that the gas has its maximal conductivity (and ionization) in this order of magnitude of the pressure.

The high voltage supply was operating with a current limitation (for instance at $50\,\mu A$), so that the voltage decreased down to $0.6\,kV$ at a pressure in the range of several $10\,mbar$ down to few tenths of a $mbar$. At this time, violet streamers could be seen optically (sometimes green streamers).

With further decreasing pressure only few single gas discharges occurred, which were visible (as soon as the room was beclouded) by looking into the window flange of the vacuum recipient and which were recognized as short-time peaks in the measurement of the electric current. When the pressure came down to about $10^{-3}\,mbar$ gas discharges did not occur any further. Below this pressure it is clear, that the drive of the rotor does not (dominantly) originate from ionized gas molecules.

Finally, the pressure was brought down to $6 \cdot 10^{-5}\,mbar$ with the rotor of fig.19 and to $1 \cdot 10^{-4}\,mbar$ with the rotor of Fig.18, which is sufficient to exclude gas discharges optically as well as by current measurement. A check of the reliability of the exclusion of gas discharges was done by enhancing the voltage. Beyond a critical voltage (approximately $17\,kV$ at the $58\,mm$-rotor) gas discharge began, visible clearly and easily under optical control as well as with current measurement.

Under full air pressure, the rotation was observed as usual (described above), but during the pumping procedure (at which electrical breakthrough and ionization occurred), no rotation could be observed any further (the voltage was too low). Even though gas ionization occurred at this pressure very much, it did not drive the rotor. It should be mentioned that the floating bodies as well as the oil degassed vehemently at the beginning of the pumping procedure, so that lots of bubbles have been produced in the vacuum oil. (There have been especially many bubbles at the $58mm$ -rotor, when the blue floating bodies have been replaced by a Styrofoam disc as floating body, because Styrofoam is not a material compatible with vacuum because of its outgassing. This material was only used in order to maximize the buoyant force of the floating body in order to minimize the friction between the floating body and the tough oil.) When degassing decreased during time, the production of bubbles decreased, coinciding with low pressure.

Nevertheless the voltage was not switched off completely during the pumping procedure and the time of gas discharge in order to maintain the self-centering mechanism of the rotor within the flux lines of the electrical field of the field source. This was necessary to avoid an adhesion of the floating bodies at the walls of the vacuum chamber due to the oil. When the pressure decreased down to values at which gas ionization did not occur remarkably, the ionization current disappeared to a value smaller than the current measurable with the present amperemeter ($1\mu A$ per each $10kV$ high voltage), so that the voltage came back to its high value (for instance to $16kV$ with regard to the limit of $17kV$ mentioned above, from which a certain distance was kept in order to avoid electrical breakthrough).

Together with the voltage, the rotation of the electrostatic rotor recurred. For example the $58mm$-rotor, with a voltage of $16kV$ and a distance of about $19\ldots20mm$ between the upper edges of the rotor blades and the field source, produced an angular velocity of one turn per 2 ... 3 hours. Please recognize these values from the pre-tests in air. But please also remember, that the voltage was smaller during the pre-tests in air. This demonstrates that both forces (a.) and (b.) mentioned above occur as long as the air is present, namely the Coulomb-force in connection with conversion of vacuum-energy as well as the force from the recoil of gas ions. When the last-mentioned force was omitted due to the removal of the gas molecules, the Coulomb-force could be enhanced by enhancing the voltage (and the field strength of the electrostatic field) enough that the rotation was still observed.

This is the central proof that that the conversion of vacuum-energy into classical mechanical energy with the electrostatic rotor introduced by the present work, really occurs, with no gas discharge in the presence of the rotor.

A short look to the voltage necessary to obtain a rotation allows a rough first estimation of the relation between the forces of (a.) Coulomb-forces and (b.) ion recoils on the basis of the produced torque for the example of the performed experiment. Therefore, please have a look to the following facts:

▪ In section 4.3 two tests with the same angular velocity are reported (this is one turn in approx 2 ... 3 hours) in air and in the vacuum, but the voltage for the rotation in air was only $10kV$, whereas the voltage for the rotation in vacuum was $16kV$.

▪ From section 4.2 the proportionality between the driving torque M and the voltage U is known, namely $M \propto U^2$.

With the notation

$M_{A,L} =$ torque according to the mechanism (a.) in air,

$M_{B,L} =$ torque according to the mechanism (b.) in air,

$M_{A,V} =$ torque according to the mechanism (a.) in the vacuum,

$M_{B,V} =$ torque according to the mechanism (b.) in the vacuum (by principle it is $M_{B,V} = 0$),

we can deduce a relation between the driving forces as following:

(i.) $M_{A,L} + M_{B,L} = M_{A,V}$ because of the equality of the angular velocity.

(ii.) $M_{A,V} = \left(\frac{16}{10}\right)^2 \cdot M_{A,L}$ because of the proportionality $M \propto U^2$.

Two equations with three unknown values can be solved to find one relation between two values. Therefore we put (ii.) into (i.) and we get:

$$M_{A,L} + M_{B,L} = M_{A,V} = \left(\frac{16}{10}\right)^2 \cdot M_{A,L} \quad \Rightarrow \quad M_{B,L} = \left[\left(\frac{16}{10}\right)^2 - 1\right] \cdot M_{A,L} = 1.56 \cdot M_{A,L} \qquad (1.1)$$

This means that for one given rotor and constant geometry (not to mix with a change of the geometry as shown in fig.15) the torque $M_{B,L}$ due to recoils of ionized gas molecules (of due to some other unknown classical mechanism) is larger than the torque $M_{A,L}$ due to Coulomb-forces on the basis of the conversion of vacuum-energy. We see now, that the recoils of ions is existing (and even dominant under air), but the main awareness of the experiment under vacuum is, that the Coulomb-forces due to the conversion of vacuum-energy really exist. And this is an important result of this experiment. Further experiments and even stronger results are following in the next sections.

4.4. "Over-unity" criterion for the exclusion of artefacts

Repeatedly the argument was heard, that a final proof of the conversion of vacuum-energy (a final verification that the rotation of the rotor is really due to the conversion of vacuum-energy and nothing else), is only give for sure, if the produced mechanical engine power is less than the electrical power losses, which occur because of imperfections of the electrical isolation due to the maintenance of the electric field [Kah 08] within the machine converting vacuum-energy. Only if this comparison of power (electrical and mechanical) is clear, experimental artefacts can be excluded for sure. The background of this argument is, to exclude all imaginable paths of energy transport from electrical energy into mechanical energy, in order to be sure, that the mechanical energy is really coming from the vacuum, because the mechanical energy could not come from anywhere else. And this is only verified with absolute certainty by the power comparison. Let us denominate this argument the "Criterion of net power gain" or the criterion of "over-unity" as some people do. The successful verification of this criterion is the topic of section 4.4.

In fact this proof was brought with a vacuum-energy rotor with an electrical power loss of $P_{el} = (2.87 \pm 0.89) \cdot 10^{-9} Watt$ in comparison with a produced mechanical engine power of approx. $P_{mech} \approx (1.5 \pm 0.5) \cdot 10^{-7} Watt$, so that at least the difference of $P_{mech} - P_{el}$ is taken from vacuum-energy for sure [e17].

As observed in section 4.3, especially in (1.68) with comment, gas ions transport electrical charge (even if not all ions produce a driving torque), so they produce an electric current, which is just an electrical power loss and thus has to be avoided for our power measurement. Indeed the electrical power loss under air is larger then the produced mechanical power. Because of this reason, the verification of the "Criterion of net power gain" has to done in vacuum [Kna 08/09]. In order to avoid the problems with a rigidly fixed axis of rotation (as it will be explained in section 5.2), which makes the operation of the rotor difficult (especially inside a metallic chamber), again a rotor swimming on oil was used. For practical reasons the floating body was made differently then those of fig.18 and fig.19. Here it was swimming like a "little boat" (see fig.21), so that are no problems with degassing during the pumping procedure for the evacuation of the vacuum chamber. By this means, the self-adjustment mechanism of the swimming rotor could be used successfully. The fluid was the same as used in section 4.3., namely the vacuum oil „Ilmvac, LABOVAC-12S". Because of the very low mass of the floating body (relatively to its volume), the immersion depth of this rotor into the oil was very small in order to avoid large friction. Nevertheless, the rotor was spinning very slowly, which can be understood with regard to the small driving torque in relation to the large viscosity of the oil.

Fig. 21: Rotor in a thin-walled plastic floating body as a lightweight construction (walls: 230µm thick polypropylene, rotor blades: 70 µm thick aluminium), on vacuum oil in an oil tub made of conducing metal. The rotor blades are connected electrically to ground with a copper filament whose end could slide on the bottom of the oil tub. The total diameter of the rotor tub is 64 mm. The oil container has an outer diameter of 97 mm.

Sharp metallic edges are protected and rounded by an isolating sealing in order to prevent the emission of electrically charged particles due to large electric field strength at the edges (which has been especially important at pre-tests under air).

Rotor and field-source have been put together with a cylindrical oil container of a diameter of 9.7 cm into a vacuum chamber with an inner diameter of a bit more than 10 cm as can be seen in fig.22.

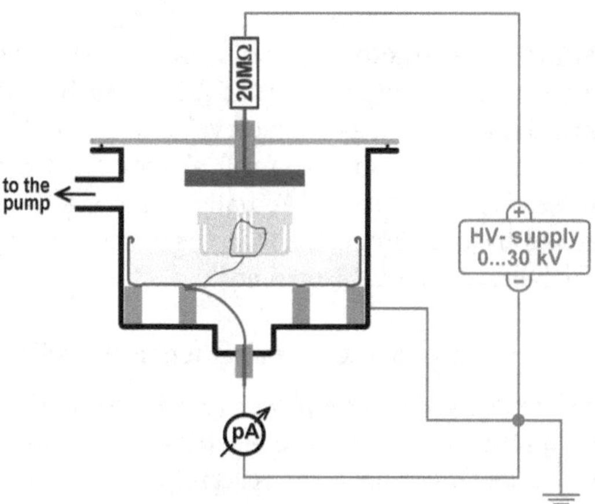

Fig. 22: Electrostatic rotor in a vacuum chamber with hydrostatic bearing swimming on oil in order to allow the self-adjustment mechanism of the rotor.
Electrically conducting components are drawn in blue, ceramic-isolators in red.
The pressure during the measurements was $4...5 \cdot 10^{-4}\,mbar$ ***, which was sufficeent for the verification of the "over-unity" criterion as will be seen soon at the evaluation of the measured data.***

Inside the vacuum chamber (made of high-grade steel, drawn in black colour, the acrylic glass of the top flange drawn in violet colour) there is the aluminium tub (blue) with vacuum oil (yellow). The polypropylene floating body (green) is swimming on the oil, carrying the aluminium rotor, which consists of four rotor blades (light blue). The rotor blades are electrically connected by copper filaments (dark blue, diameter 30 µm) with each another and with the bottom of the oil container. The oil container and the rotor are electrically insulated with ceramic insulators (red) against the vacuum chamber. Every electrical charge taken up by the rotor must flow through the Picoamperemeter Keithley 486 („pA ") and is detected there. With a well-known value of the high voltage (grey, equipment „Bertan ARB 30 ") the electrical power taken by the rotor is determined (if there is such power at all).

The electrostatic field for the drive of the rotor is produced by the field-source made of aluminium (dark blue), which is mounted at the top flange held by a ceramic isolator. In order to avoid a damage of the Picoamperemeter in the case of electric breakthrough, a resistor of 20 Megaohms is put between the electric power supply and the field-source. In the real measurement 20 Megaohms is such small resistance (in comparison with the disconnection between field-source and rotor), that it will not be noticed at all.

Only if the experimentation parameters are adjusted properly (which is not yet under complete control and thus has to be found by trial and error), the rotor begins to spin as soon as the high voltage is switched on – if the torque which the rotor takes up inside the electric field is strong enough to surmount the friction of the oil. Actually, there is a threshold, which must be exceeded by the torque, so that the rotor can begin to spin. (Obviously, the oil behaves like a thixotropic fluid.)

As soon as the rotor spins, the goal of the measurement is the following:

On the one hand, the mechanical engine power (produced by the rotor) has to be determined; on the other hand, the electrical power loss has to be determined, which is inevitable in a real existing setup because of practical reasons. At least some very small creepage or leakage currents result in a loss of electrical charge on the field-source, and this amount of charge being lost has to be replaced in order to keep the electric field constant. Of course, in the case of ideal isolation, no electric charge would be lost and thus no electrical power loss would arise. But in reality, at least some atoms of the residual gas inside the vacuum chamber will produce some loss of electrical charge. A successful proof of the conversion of vacuum-energy is given, if the electrical power loss is clearly smaller than the mechanical power driving the rotor, because in this case the electrical power can not be sufficient to explain the spinning of the rotor. This is a clear definition of the goal: It consists of two measurements, namely the mechanical power and the electrical power.

Same as in the practical performance of the experiment, also the report here shall begin with the mechanical power, because from its knowledge we see the requirements for the electrical power measurement.

Part 1: Measurement of the mechanical engine power

The mechanical power has been measured „ex-situ" outside the vacuum chamber, with the rotor swimming on the oil within the oil tub, which was located at the bottom of a rack as shown in fig.23. The driving torque was produced by a thin copper torsion-filament (diameter $50\,\mu m$) being twisted by a well defined angle, thus producing a well defined torque in the following way: At first it was waited until the came to its rest position, this is a torque of zero. Than the torsion-filament was twisted at its top end with a hand wheel, so that it produces a well defined torque making the rotor spin. From the knowledge of the torque and the angular velocity of the rotor, the mechanical engine power driving the rotor could be determined as shown below. With the means of such measurements, a mathematical connection between the duration for one turn and driving engine power was determined for several values of angular velocity.

And this is the procedure how to determine the measurement of the mechanical engine power:

<u>Step 1:</u> Characterization of the copper filament (torque versus the angle of torsion)

As a preliminary work it was necessary to determine the torque of the filament as a function of the angle of distortion. For this purpose the rotor at the bottom end of the filament was replaced by a hollow plastic sphere (diameter $r_a = (39.7 \pm 0.1) \cdot 10^{-3} m$, mass $m = (2.732 \pm 0.002) \cdot 10^{-3} kg$) not using any oil in this phase of the experiment. Now the rotating wheel on the top was deflected, resulting in an oscillation of the sphere (duration of one period $T = (19.76 \pm 0.02)$ sec., length of the filament $l = (409 \pm 1) \cdot 10^{-3} m$). With elementary formulas of technical mechanics [Dub 90], [Tur 07], the mathematical expression $Q = \frac{G \cdot \pi \cdot R^4}{2 \cdot l} = (2.902 \pm 0.016) \cdot 10^{-7} Nm$ was calculated (G = shear modulus of the copper and R = radius of the filament), which is a factor of proportionality between the torque and the angle φ of the torsion, namely $M = Q \cdot \varphi$. This is a calibration of the copper torsion-filament.

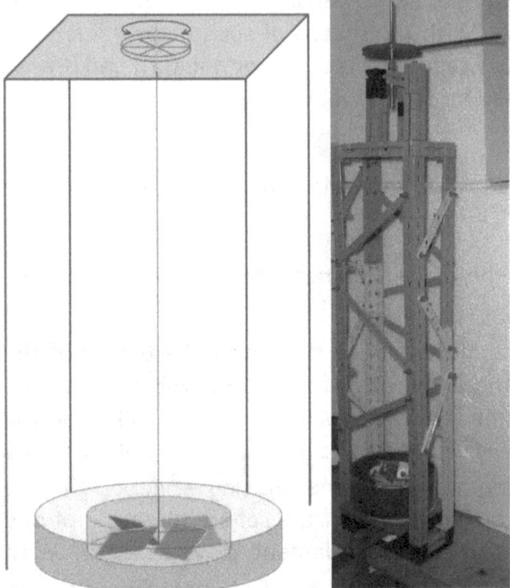

Fig. 23: Sketch and Photo of the setup for the determination of the classical mechanical power as a function of the duration per turn.
The rotor is fixed at the bottom end of a copper torsion-filament (purple) with which a torque can be applied to the rotor.
It was paid attention to the immersion depth of the rotor inside the oil which has to be (as good as possible) the same as in the vacuum chamber in order to get the same friction between the oil and the rotor in both cases.

Step 2: Determination of the rotor's moment of inertia of rotation

Because of the irregular shape of the rotor with its rotor blades, it is not sensible to calculate the moment of inertia theoretically, but it is more accurate to measure it. Therefore, the hollow plastic sphere at the bottom end of the filament was replaced by the complete rotor with the floating body, still not using any oil. Now the rotating wheel on the top end of the filament was deflected again, resulting in an oscillation with a duration of one period of $T_b = (33.70 \pm 0.06)\,\mathrm{sec.}$ at a filament length of $l_2 = (383 \pm 2) \cdot 10^{-3}\, m$, leading to an expression of $Q_2 = \frac{G \cdot \pi \cdot R^4}{2 \cdot l_2} = (3.099 \pm 0.025) \cdot 10^{-7}\, Nm$. The torque at this filament length is now $M_2 = Q_2 \cdot \varphi$. With standard formulas of technical mechanics we come to moment of inertia of $Y = \frac{Q_2 \cdot T_b^2}{4 \cdot \pi^2} = (8.916 \pm 0.078) \cdot 10^{-6}\, kg \cdot m^2$ for the rotor together with its floating body.

At this phase of the experiment, the torsion-filament is characterized as well as the rotor together with its floating body, as far as it is necessary for the determination of the mechanical engine power as a function of the angular velocity of the rotor on the vacuum oil. This function will be determined now:

<u>Step 3:</u> Determination of the mechanical engine power as a function of the duration per turn

Because of practical reasons, namely because of the proper adjustment of the rotor on the surface of the oil, the length of the torsion-filament had to be changed once more, namely to $l_3 = (420 \pm 2) \cdot 10^{-3}\, m$, corresponding with a value of $Q_3 = (2.826 \pm 0.022) \cdot 10^{-7}\, Nm$.

Now the rotor could easily be driven by different well known values of the torque by turning the rotating wheel for a given angle φ because of the proportionality $M_3 = Q_3 \cdot \varphi$. From the torque M and the angular velocity ω_u, the engine power $P = M \cdot \omega_u = Q_3 \cdot \varphi \cdot \frac{2\pi}{T_\alpha}$ is calculated. The result is plotted in fig.24, where we see the engine power as a function of the duration T_α of a single turn. The engine power at a duration in the range of $\frac{1}{2} ... 1 ... 2\ hours$ should be memorized, because this is, what we will observe later at the rotor inside the vacuum.

The curve in fig.24 also contains an aspect of preparation of the electrical power measurement in part 2 of the measurements (following after fig.24), because it tells which voltage and which current we have to measure for the determination of the electrical power. This can be demonstrated by the following estimation with exemplary values:

Let us assume a voltage of $U \approx 25 kV$ to make the rotor spin with one turn per one hour, so fig.24 leads us to a mechanical power in the range of approx. $P_{mech} \approx 1.5 \cdot 10^{-7}\, Watt$. If the electrical power loss shall be remarkably smaller than the mechanical power, the electrical current has to be $I \ll \frac{P}{U} \approx \frac{1.5 \cdot 10^{-7}\, Watt}{25 \cdot 10^{3}\, Volt} = 6 \cdot 10^{-12}\, Ampere$. This is the demand to the measurement of the current.

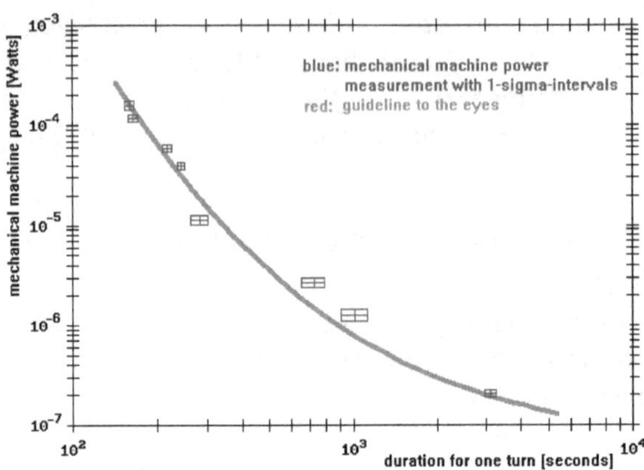

Fig. 24: Mechanical engine power of the rotor inside the floating body when rotating on the surface of the vacuum-oil.
A duration of one turn per $\frac{1}{2}...1...2\ Stunden$ ***leads to a mechanical power of about*** $1...2\cdot 10^{-7}\,Watt$.

Part 2: Measurement of the electrical power loss

This part of the measurements was done in the vacuum as said above.

For the purpose of comparison, a measurement with an empty vacuum recipient was done first, without high voltage, without oil, without rotor. The values of the electrical current have been recorded automatically (electronically) in order to be available for a later computation of the integral average value. With a duration of measurement of 30 seconds, we come to an integral average current of $I_1 = (0.08 \pm 0.01)\,pA$. This demonstrates the limit of the resolution and of the precision of the current measurement, which fits the requirements known from the comments to fig.24.

After putting the rotor on the oil in the oil tub into the vacuum recipient (before closing the top flange of the chamber), it has been checked that the rotor blades and the negative contact of the high voltage supply are connected properly as well as the fact that the field-source and the positive contact of the high voltage supply are connected properly. In the same way it was checked, that (still without the Picoamperemeter being connected and without grounding the vacuum chamber) the vacuum chamber, the field-source and the oil tub with the rotor are electrically disconnected (test with a Mega-Ohmmeter, Fluke). In reality, a very very small current between the field-source and the rotor is possible, as we can see later after analyzing the data, but the resistance is by several orders of magnitude too large for a detection with our Mega-Ohmmeter, and the leakage currents are small enough that they will not disturb the aim of our measurements. This indicates, that also the atoms of the residual gas in vacuum chamber (when being evacuated later) will transport less enough electrical charge, that the electrical power loss does not prevent the aim of our measurement. From this observation we can conclude that it is not necessary to improve the vacuum to a value better than given in fig. 22. To reduce the pressure even more would have been a problem because of the presence of the oil.

After the pre-tests of the electrical connections were finished, the top flange was closed, carrying the field-source. Now the high voltage supply was to be connected, but not the Picoamperemeter (still without evacuation of the chamber), so that and electrical field could be produced which activates the self-adjusting-mechanism of the rotor. Only after observing a rotation of the rotor, the vacuum pumps have been switched on. This rotation presumes substantial precision work to adjust all experimental parameters in an appropriate way, as for instance the amount of oil inside the tub, the distance between the rotor and the field-source, an adequate value of the voltage and so on... The coordination of experimental parameters has to be balanced properly, which was done by trial and error and is based on some practical experience in the laboratory.

Now the vacuum pumps have been switched on. At first we see a degas procedure of the vacuum oil, which is rather strong, because gas bubbles can elude only very slow from the oil because of the rather large viscosity of the oil. It is necessary to have a current limitation at the high voltage supply, because the residual gas will be conducting when the pressure passes the range of few millibars, where gas- and corona- discharges can be observed easily by optical visible luminous effects. The current limitation of the high voltage supply prevents the apparatus from being damaged. But on the other hand the high voltage supply can not be switched off completely during the beginning of the pumping procedure, because without any high voltage, the gas bubbles from the degas procedure of the oil might move the rotor to the side wall of the oil tube, where it would adhere because of the toughness of the oil. Only rather seldom, it is possible to get the rotor when sticking to the side wall of the oil tub away from this position just only by using the self-adjustment mechanism. In the most cases, a rotor sticking to the side wall of the oil tub can only be removed after aerating the vacuum recipient and beginning the whole procedure of the electric part of the measurement from the very beginning.

During the further evacuation of the chamber, the conductivity of the residual gas decreases (together with the visible gas- and corona-discharges), so that the necessity for the current limitation of the power supply will not exist any further. This means that the high voltage can be regulated from 0 to 30kV, according to the specification of the power supply and following the requirements of the rotor.

Only when the degas procedure of the oil is mostly done (visible from the number of gas bubbles in the oil), the pressure in the vacuum recipient will come down to the value of $4...5 \cdot 10^{-4} mbar$ as said in fig.22. Now (after switching off the high voltage, which is now possible, because there are no more gas bubbles moving the rotor to the side wall of the oil container) all electrical connections are checked again and than all cables can be connected according to fig.22, also the Picoamperemeter, because the risk for large electric currents does not exist any further. At this phase of the experiment, the electrical cabeling is complete.

This is the moment to enhance the high voltage beginning from $0\,Volt$ to a value that the rotor will again begin to rotate. If there are still some last gas bubbles in the oil, they will enhance the noise of the electrical signal of the current measurement (independently from the fact whether the voltage is high enough to make the rotor spin or not), which probably has is reason in the fact, that the bubbles move the rotor vertically, changing its distance from the field-source permanently back and forth. Because this electrical noise is rather strong, we had to wait with the measurement of the electrical current until this type of noise was over.

The voltage, which is necessary to make the rotor rotate, is higher than the voltage, which activates the self-adjusting-mechanism (except for the case that the rotor adheres at the side wall of the oil tub). Depending on the distance between the rotor and the field-source, the voltage for the self-adjusting-mechanism could be between 3 and 20 kV and the voltage for the rotation could be 5 and 30 kV. But these are not exact values like a result of the measurement. These values shall only give a feeling in which range the rotor is operated. If the rotor is sticking to the side wall of the oil tub, in some cases (not too often) it can be brought back in its working position by applying 30 kV, in order to produce a very strong force of the self-adjustment mechanism. But this includes the risk that the rotor is lifted out of the oil and will be flying directly towards the field source until it touches the field source.

With the rotor rotating, the voltage is measured and the current is recorded electronically again in order to get the data for the determination of the integral average value of the current later. The voltage was kept constant by the power supply. The current is statistically noisy with amplitudes up to some picoamperes, but with alternating algebraic sign, indicating the alternating direction of the current. This means that the electrical charges move statistically back and forth, not bringing electrical power into the rotor. This is the reason for the computation of the integral average value, because the charges going back and forth do not supply the rotor with power. (This is a DC-experiment.) But after evaluation of the data, the integral average value turned out to be not completely zero. This integral average value of the electrical current has to multiplied with the voltage in order to determine the electrical DC-power existing, which has to be compared with the produced mechanical power.

Data of a practical measurement: At a voltage of $U = 29.7\,kV$, the rotor rotated with a duration of about (1±½) hours for one turn. The integral average value of the current (duration of the measurement 90 seconds) was $I_3 = (-0.100 \pm 0.030)\,pA$. The noise and with it the uncertainty of the integral average value is larger than it was without rotor, although the duration for the averaging was larger. The uncertainty of the integral average value is given as 1-sigma-intervall.

The algebraic sign of the electric current is not interpreted at all. It only indicates the direction in which the electric current passes the Picoamperemeter and thus it does not have any importance for the determination of the electric power loss.

The electrical power loss is

$$P = U \cdot I = 29.7 \cdot 10^3 V \cdot (0.100 \pm 0.030) \cdot 10^{-12} A = (2.97 \pm 0.89) \cdot 10^{-9}\,Watt\,.$$

It might be due to imperfect isolation, but its reason is not really important, because it is much smaller than the produced mechanical power, which is only

$$P_{mech} = (1.5 \pm 0.5) \cdot 10^{-7}\,Watt\,.$$

For the sake of illustration, the result is plotted into the diagram of the power balance, coming from fig.24 to fig.25.

As can be seen, the electrical power loss is by about one and a half orders of magnitude smaller (this is the order of magnitude of a factor of 30) than the produced mechanical power. Whatever reason the power loss has – one fact is obvious: The rotation of the rotor can not be explained by the electrical power. And because there is no other way, along which any energy and any power could be delivered to the rotor, the rotor can only be driven by vacuum-energy. This is the result of the presented experiment.

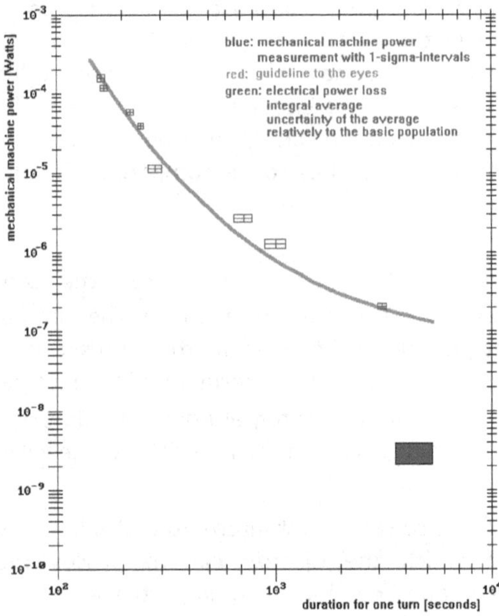

Fig. 25: Comparison of the electrical power loss with the produced mechanical power. The uncertainties are given as 1-sigma-intervalls (distribution of Gauss).

Annotations and discussion of the results

The presented experimental setup was built up with uncomplicated technology, using material, which was available in the laboratory as far as even possible. Effort and expenses for complicated equipment was only spent for those components, for which such effort was absolutely inevitable because of functional reasons (as for instance for the vacuum recipient attached to a turbo molecular pump or for the measurement of the electric current with a precision of $10^{-14}\,Amperes$). This approach allowed to test the quoted theory of the conversion of vacuum-energy very quick (the experimental part of the work was done within a bit less than one and a half years), as it would not have been possible with long preliminary lead time such as it would have been necessary for the application of financial aid, for complicated constructions, for buying complicated devices, and so on.

The fact, that the result is so clear and unambiguous despite the simplicity of the setup (without the artefacts possibly arising from complicated machines), confirms the quoted theory of the conversion of vacuum-energy into classical mechanical energy in a strong way.

Such a clear result encourages to perform now further developments and investigation, to put more effort into further experiments with the goal to get all experimental parameters under clear control (instead of the trial and error which could not be avoided in several cases up to now). If such a detailed knowledge of the machine for the conversion of vacuum-energy could be developed, it should also be possible to enhance the mechanical engine power being converted from the vacuum-energy. It gives hope that it should be possible to enhance the mechanical power to a range, which is capable for practical applications. This hope arises from the proportionality, which says, that the engine power is being enhanced with the square of the diameter of the rotor. This enhancement of the mechanical power might be accompanied by a reduction of the gas pressure in the vacuum, so that the electrical power loss will be even reduced.

The physical principle of the presented energy source is proven with the present work, but now the application as an energy source is visible, which might allow to get some of the vacuum-energy out of the space of the universe – if it is possible to enhance the converted power enough.

The clearness of the results presented here should hopefully inspire other scientific groups of fundamental experimental physics and of engineering sciences to reproduce the experiment, and first of all, to optimize it. Of course, there will be several experimental difficulties to be overcome, but it looks worth being done. In section 5 there will be some consideration regarding future experimental difficulties and optimizations.

5. Outlook to the future

5.1. Magnetic analogue with the electrostatic rotor

With regard to the similarities between the circulation of energy of the electric field and the circulation of energy of the magnetic field, as demonstrated in section 2.3, it is expected, that it should be possible to build a magnetic rotor for the conversion of vacuum-energy in analogy with the electrostatic rotor for the conversion of vacuum-energy [e15]. A suggestion for an appropriate setup is conceived in section 5.1. The considerations also contain exemplary calculations of the forces being expected (in comparison with the force being expected from the electrostatic rotor).

In order to understand the rotation of the electrostatic rotor, the Coulomb-force of an electric field acting onto a rotor of electrically conducting material was calculated with the image-charge method [Bec 73]. One central fundament for the applicability of this method is the fact, that the electric flux coming from outside onto the surface of the material rearranges the electrical charge distribution on the surface in such a way, that the flux-lines are always exactly perpendicular to the conducting surface [Jac 81] and the field does not penetrate into the inside of the material. The force between the charge and the image-charge explains the rotation of the rotor-blades and thus it shows the instrument to get some energy out of the cycle of energy between electrostatic energy and vacuum-energy.

If this principle of energy conversion shall be adopted to the energy cycle between the magnetic field energy and the vacuum-energy, a rotor driven by magnetic forces should be constructed, which can be understood in analogy to the rotor driven by electrostatic forces. The analogue to the image-charge method is to be found in the Meissner-Ochsenfeld-effect occurring on the surface of a superconductor in a magnetic field [Tip 03]. Similar as the image-charge method is based on the appearance of electrostatic charges on the surface, the Meissner-Ochsenfeld-effect is based on the appearance of superconducting currents on the surface of a superconductor in a magnetic field. These currents evoke magnetic fields, which exactly compensate the external magnetic field, so that in the inside of the superconductor the total magnetic field is zero. (In the inside of the electrical conductor the electrical field is also zero.) In this sense, the superconductor displays ideal diamagnetic behavior with a susceptibility of $\chi=-1$ [Ber 05].

In order to calculate the forces which superconducting rotor-blades take up in a magnetic field, the field strength at each point on the surface of the rotor-blades have to be taken into account and their interaction with the field strength of the same absolute value but of the opposite orientation has to be determined (finally going back to the associated electric currents). An exemplary setup of an imaginable rotor is show in fig.26 with a permanent magnet in the above part of the sketch and a rotor consisting of four superconducting blades in the bottom part of the sketch. The geometrical dimensions noted there serve as input parameters of an exemplary calculation, which has the purpose to give a feeling of concretely achievable forces and torques. The forces are repulsive (not attractive), because the flux-lines can not penetrate into the ideal diamagnetic material, independently from the orientation of the external magnetic field produced by the permanent magnet. It is generally known that the Meissner-Ochsenfeld-effect is independent from the polarity of the flat permanent magnet. Thus, the direction of the rotation of the rotor in fig.26 is counterclockwise (other than the direction of rotation of a similar electrostatic rotor). The electric forces between charges and image-charges are attractive, but the magnetic forces between currents and "image-currents" are repulsive.

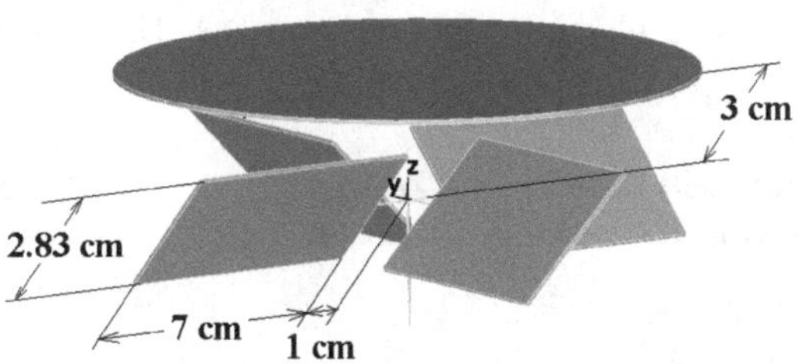

Fig. 26: Graphical visualization of a superconducting rotor, which is mounted on an axis below a flat permanent magnet.

Calculation of a numerical example

For the sake of simplicity of a numerical example, the external magnetic field driving the rotor shall be premised to be homogeneous (with a numerical value of $|\vec{H}|=1\frac{A}{m}$), and it shall be orientated parallel to the z-axis.

Thus, the vector of the magnetic field is the same on all positions on the surface of the rotor-blades. Consequently, the magnetic field produced by the Meissner-Ochsenfeld-effect (which compensates the external field) is orientated exactly into the negative z-direction for all positions of the rotor-blades (and has the absolute value of $|\vec{H}| = 1\frac{A}{m}$). But the position-vectors from all individual points of the field source to all individual points on the surface of the rotor-blades are going into individually different directions. And each rotor-blade is not symmetric with regard to the z-axis. Consequently, the total force acting onto each rotor-blade, which is the sum of all forces acting between each individual part of the rotor-blade and each individual part of the field source, contains a tangential component, which is not orientated into the z-direction. This consideration implies a subdivision of the field source and the rotor-blade into finite elements.

The calculation of the sum of all finite force elements is done as following: Each finite element of the field source produces a field at each finite element of the rotor (this field element produced by the magnet shall be numbered with the index no.1). And the finite element of the rotor itself reacts to the total field with its orientation into the opposite direction (the finite field reaction element of the superconductor shall be numbered with the index no.2). The magnetic forces between all these pairs of field elements are determined as usual on the basis of Biot-Savart's law and the law of the Lorentz-force. But here these formulae are written in a way, that we directly see the forces between the two field strengths of the particular field element pairs ($\vec{H}_1$ and $\vec{H}_2$). This is the following formula:

$$\vec{F}_{12} = 4\pi\mu_0 \cdot |\vec{r}_{12}|^2 \cdot (\vec{e}_{12} \times \vec{H}_1) \times \vec{H}_2 \ ,$$

$$\text{with } \mu_0 = 4\pi \cdot 10^{-7} \tfrac{Vs}{Am},\ \vec{r}_{12} = \text{vector connecting the field-elements} \qquad (1.69)$$

$$\text{and} \qquad \vec{e}_{12} = \text{unit-vector between the field-elements.}$$

Because the field strength has been chosen to be homogeneous in our example (absolute values $H_1 = -H_2 = 1\frac{A}{m}$), the total force onto a rotor-blade can be determined by a rather simple integration over the surfaces of the field source and the blade (keeping in mind the double vector-product for each field-element-pair). For the given example, this was carried out by a subdivision of the blade as well as the field source into finite elements and summation of all these force-elements. If the subdivision is refined more an more (with growing number of elements), the result (on the basis of the given numerical parameters) converges to a force values of $F_R = 4.2 \cdot 10^{-8} N$ for the radial component of the force and $F_T = 3.1 \cdot 10^{-10} N$ for the tangential component of the force per each single rotor blade. The radial component of the force is compensated by the mechanical axis of the rotor, whereas the tangential component of the force should make the rotor spin. A rotor consisting of four blades takes up four times the tangential force, this is $F_T = 1.2 \cdot 10^{-9} N$. The torque onto the rotor was calculated by summing up the torque of each finite element pair of the rotor with its individual radius of rotation, which leads to the result of $M = 4.8 \cdot 10^{-11} Nm$.

Please be aware, that the radial component absorbed by the bearing of the axis of rotation is the dominant part of the force. This leads to the consequence that the experimental setup has to be built with extreme mechanical precision, otherwise a small part of the radial force component could already disturb the rotation. Additionally the homogeneity of the magnetic field has to be very good (better than a single pair of permanent magnets can do) and the mechanical adjustment has to be done with an adequate precision. Otherwise the rotor would not spin.

The calculated value of the torque is too small for an experimental verification, but the field strength can be enhanced remarkably and with it the forces and the torque. They increase quadratically with the field strength according to (1.69), following the proportionality of the absolute values $|\vec{M}| \propto |\vec{H}_2| \cdot |\vec{H}_1|$. The limit of the field strength is given by the critical field strength B_c of the Meissner-Ochsenfeld-effect, which depends on the temperature. For instance, a superconductor type 1 can have $B_c = 0.080 T$ for Pb or $B_c = 0.01 T$ for Al (with temperature extrapolated $T \to 0 K$). In order to have some safety distance from the critical field, and in order to respect that the fact the temperature in a real experiment is different from zero, we can allow a magnetic induction B of few milli Tesla, corresponding with a field strength of about $H_1 \approx 10^3 \frac{A}{m}$. This lets us expect a the tan gential component of the force in the order of magnitude of $F_T \approx 1.2 \cdot 10^{-9} N \cdot 10^6 \approx 1.2 \cdot 10^{-3} N$, providing a driving torque of about $M \approx 4.8 \cdot 10^{-5} Nm$. This should be sufficient for a measurement, namely to surmount the friction of a real existing bearing as said in (1.65), (1.66), or (1.67).

If cooling should be made as easy as possible by using liquid nitrogen, it could be interesting to work with a high-temperature superconductor, as for instance $YBa_2Cu_3O_7$, which is known to be a superconductor type 2. It only behaves ideal diamagnetic as long as the Shubnikov-phase is avoided. Nevertheless the field strength and with it the maximum of the reachable forces and the torque should be expected to be even larger than with the superconductor type 1 as mentioned above. In this sense, a conversion of vacuum-energy via magnetic field energy into classical mechanical energy should be detectable even under liquid nitrogen.

Continuative remarks regarding an experiment

Another question is, whether the practical performance of the experiment really needs ideal diamagnetic rotor-blades made from superconductor material. Also conventional dia-, para-, and ferromagnetic materials experience forces when they are exposed to magnetic fields even at room temperature. This question can not be answered here, but it is possible to develop some thoughts about forces and torques.

If a rotor would be made from classical diamagnetic metal (at room temperature), for instance such as copper ($\chi = -1 \cdot 10^{-5}$) or bismuth ($\chi = -1.5 \cdot 10^{-4}$) (see [Stö 07]), it would be $H_2 = -\chi \cdot H_1$. With the same strength of the external field, the forces and the torque would be about 4 or 5 orders of magnitude smaller compared to a superconducting rotor, but the field strength could be enhanced perhaps to 1 Tesla of even more rather easily, so that finally the torque might be enhanced to a value even a bit larger than with a superconducting rotor. The mentioned values lead us to a torque in the order of magnitude of about $10^{-5} \ldots 10^{-4}\, Nm$.

Instead of using a diamagnetic rotor, it might be discussed to build up a paramagnetic rotor, for instance of platinum ($\chi = +1.9 \cdot 10^{-6}$) or aluminium ($\chi = +2.5 \cdot 10^{-4}$) (see [Ger 95]). The forces will be attractive in this case so that the rotor spins with opposite direction than a diamagnetic rotor (if it spins), but the absolute values of the torque should be of the same order of magnitude as those of diamagnetic rotor.

Large forces might be expected from a ferromagnetic rotor. If it would be sensible to ascribe a susceptibility to ferromagnetic materials, this could lead for some materials to values several orders of magnitude larger the susceptibility of dia- or paramagnetic materials. But ferromagnetic materials produce a magnetization because of a regularity of the electron spins, which depends on the previous history of the material [Kne 62]. Thus, it is very doubtful whether it is allowed to carry forward our theoretical considerations to ferromagnetic materials at all. The image-charge method had been applied to the electrostatic rotor, and its analogue was developed for an ideal diamagnetic surface. Consequently it is not clear, whether these considerations are sufficient for dia-, para-, or ferromagnetic materials. In any case, the analogy with the currents compensating the magnetic field or the analogy with the image-charge method compensating the electric field is not clear for such materials. All the more we face this question of uncertainty regarding ferromagnetic rotor-blades, in which magnetic fields influence the whole spin-order [Kne 62]. Ferromagnetic domains are generated and kept in the material because of the Barkhausen-effect. This arises serious doubts, whether the spins and magnetic moments can follow an external magnetic field, as it is required for a rotor, which converts vacuum-energy into magnetic and mechanical energy.

Furthermore, for the practical construction of a magnetic rotor, it has to be remembered again, that the above calculation came to the result, that the tangential component of the force (causing the rotation) is about two orders of magnitude smaller than the radial component of the force (disturbing the rotation). Thus, it is essential to build up the mechanical arrangement of the components with extremely high precision in order to avoid that the radial forces will prevent the rotation. Deviations from the ideal exact movement of the rotor have to be kept such small, that the rotor will not find a minimum of energy at some position, which stops its rotation. Forces keeping the rotor in the position of an energy minimum because of tolerances in the mechanical setup have to be smaller than the driving tangential forces. This problem of a rotor being stopped in a minimum of energy regarding the tangential force was already practically experienced, but the magnetic rotor was not yet rotating in satisfactory in an experiment.

Also, the homogeneity of the magnetic field is a very critical and important aspect. This was also experienced in practical tests [Lie 08/09], and it also caused a standstill of the rotor, mostly after less then one turn, in very few cases between one and two turns, but even the angle of rotation was not reproducible. Nevertheless, the experiment for the conversion of vacuum-energy via magnetic field energy into classical mechanical energy seems worth being further developed until it will be successful.

The problem for the magnetic rotor is, that the self-adjustment mechanism (as known from the electrostatic rotor) does not exist for the magnetic rotor. But it was the self-adjustment mechanism, which made the rotation possible with a rotor of moderate mechanical precision and moderate field homogeneity. The reasons for the lack of the self-adjustment mechanism in the magnetic case is the following: The Coulomb-forces driving the electric rotor are attractive, but the Lorentz-forces driving the (superconducting) magnetic rotor are repulsive. Attractive forces try to minimize the distance between the rotor and the field source in a way that a lateral movable rotor is being transported into the minimum of the potential – this is the self-adjustment mechanism of a rotor driven by attractive forces. Repulsive forces behave in the opposite manner. They transport a lateral movable rotor away from the field source. This is the lack of the self-adjustment mechanism of the magnetic rotor. It can be avoided only by a rigid fixation of the axis of rotation. With other words: If the magnetic (superconducting) rotor shall remain within the field, it needs a fixation of its axis of rotation, and thus the self-adjustment mechanism can not work at all. This is the reason why an experiment with moderate effort and expenses (as done in section 4.4) is not possible for the magnetic rotor.

After these explanations and calculations it is clear, that it should be possible by principle to convert vacuum-energy not only via electrostatic field energy but also via magnetic field energy. But the experimental challenge for the magnetic conversion is different from the experimental challenge for the electric conversion – and rather demanding. The magnetic conversion does not require vacuum, because magnetic fields do not ionize gas atoms (other than electric fields). A measurement of energy and power might be dispensable for the magnetic conversion, if the field is produced by a permanent magnet, because a permanent magnet does not take power at all. But the demand regarding the mechanical precision and the demand regarding the homogeneity of the field is very high for the magnetic energy conversion.

By the way: If the electric conversion of vacuum-energy shall be done with rigidly fixed axis of rotation (not with the hydrostatic bearing of the swimming rotor), the demand to the mechanical precision and to the homogeneity of the field is the same high as it is for the magnetic rotor. Therefore, the following section 5.2 is written.

5.2. Rotor with rigidly fixed axis of rotation

As mentioned in the numerical example of section 5.1, there are not only the tangential forces acting onto the rotor (causing a rotation) but there are also radial forces and vertical forces which want to move the rotor by translation into horizontal of vertical direction. If a rotor as seen in fig.27 (uncomplicated to assemble) is put onto a toe bearing, the risk is rather large that it falls down, when the field is switched on. Only when the adjustment is very good, it might rotate, but it can happen, that it falls down during the beginning of the rotation. These observations indicate the transverse forces, which act onto the rotor. It is under investigation whether those transverse forces are due to the limited mechanical precision of the rotors analyzed up to now (but by principle every mechanical assembly always has limited precision) and due to inhomogeneities of the field (which are also inevitable by principle) [Bec 08/09].

It should be said, that the rotor of fig.27 has a stable position on the tip of the toe bearing if the field is not switched on, because its centre of mass is located lower then the support point. (Even when the setup is carried by hand from one table to another without special caution, the rotor does not fall down from the toe bearing.) This illustrates the extent of the transverse forces (horizontally of vertically) in the electric field (for the electrostatic rotor). In fact, it was not possible to operate the rotor on fig.27 inside the evacuated vacuum chamber without falling down – other than in air, where the gas ions give an additional propulsive force, causing a rotation, which stabilizes the rotor. Thus under air a rotation was possible (if the adjustment was good enough), sometimes with a plane of rotation not parallel to the plane of the field source. In the vacuum the rotor always fell down, when the voltage has been switched on. In this context we understand, why the transverse forces are more important in the vacuum than in the air.

Fig. 27: Uncomplicated bent rotor for the demonstration of the transverse forces, which occur when the electric field is switched on. The lateral self-adjustment mechanism is not possible because of the rigidly fixed axis of rotation, so that the transverse forces do not cause the self-adjustment of the rotor but they make the rotor fall down from the toe bearing.

Independently from the difficulties with the technical realization, it is desirable to operate rotors for the conversion of vacuum-energy with a mechanically fixed axis, because on the one hand the hydrostatic bearing with the rotor swimming on oil is nonpractical for technical application (especially under vacuum), and on the other hand the fixed axis is possible only in the case of the electrostatic rotor, not in the case of the magnetic rotor.

For these reasons, an investigation of the transverse forces acting on the rotor during its rotation is important. This was done by filming a swimming rotor (producing short little movies of 30 seconds with a digital camera) during its rotation. For this purpose the rotor of fig.21 was used once more (diameter of the rotor 64 mm). The investigation was performed under air. The rotor was put onto a water surface inside a metal tub with a diameter of 18 cm. The field source had a diameter of 13 cm and was mounted few centimetres above the rotor. The short movies have been made during the rotation of the rotor driven by several values of the voltage (between field source and rotor).

The evaluation of the data was done with regard to the angle of rotation and with regard to a lateral movement of the rotor (following the self-adjustment mechanism). The lateral position of the rotor was taken in coordinates "x" and "y" as shown in fig.28, where "x" is the outermost left point of the rotor's floating body and "y" is this point of the rotor's floating body, which is most close to the camera. The movement of these both points are to be understood as x- and y- coordinates of the rotor (and thus also as the xy-position of the rotor's axis). The investigations has been performed under air, but it is clear, that the lateral movement also occurs under vacuum. Actually, fact is that the lateral movement is stronger under vacuum then under air (as explained above).

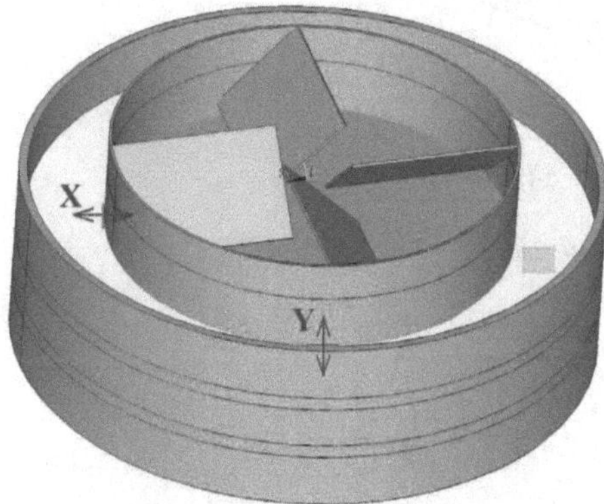

Fig. 28: Rotor with floating-body on the surface of a fluid (yellow) inside a metallic tub. The lateral movement of the rotor edges (marks at the edges in red, movements in dark blue) has been filmed. The light green rectangle symbolizes a piece of Styrofoam of well defined dimensions (1.0 cm x 1.0 cm) for the calibration of the x- and y- scale.

The calibration of the x- and y- scale was done with a small piece of Styrofoam, whose edges have been orientated parallel to the x- and y-axis. After the calibration was done, the traces of the rotor in the xy-plane (during the time of rotation) have been analyzed by counting the pixels of the x- and y- position of the digital pictures of the movies (one picture after the other). The angle of rotation was found by looking to the coloured pattern on the outside of the rotor's floating body (the rotor with floating body is shown in fig.21).

Three typical examples of such traces are plotted in fig.29 (for different voltages producing the electrostatic field).

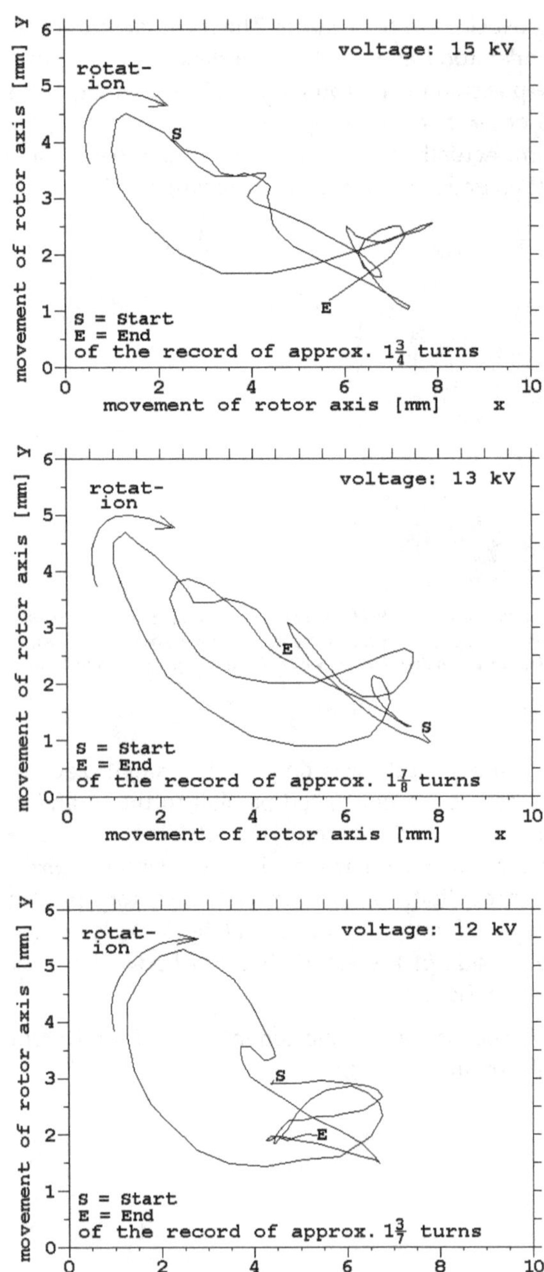
voltage: 15 kV
rotat-
ion
S
E
S = Start
E = End
of the record of approx. 1¾ turns
movement of rotor axis [mm] y
movement of rotor axis [mm] x
voltage: 13 kV
rotat-
ion
E
S
S = Start
E = End
of the record of approx. 1⅞ turns
movement of rotor axis [mm] y
movement of rotor axis [mm] x
voltage: 12 kV
rotat-
ion
S
E
S = Start
E = End
of the record of approx. 1 3/7 turns
movement of rotor axis [mm] y
movement of rotor axis [mm] x

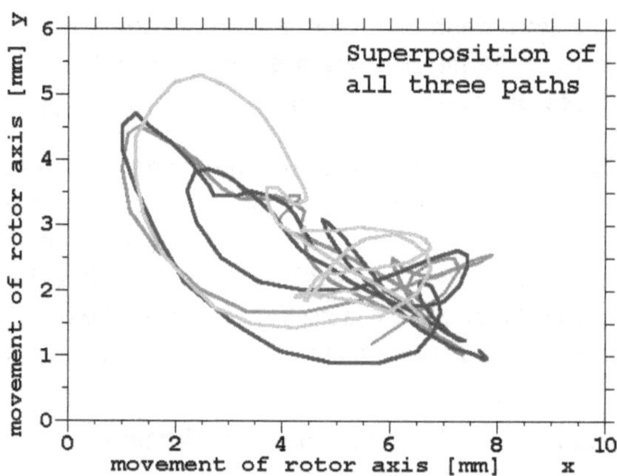

Fig.29, first part: Traced paths of the lateral movement of the swimming rotor during its rotation on a water surface. The position of the rotor is moving because of the self-adjustment mechanism. Alltogether we see three examples of paths of the rotor position, measured during the rotation at three (slightly) different values of the field strength (voltage between rotor and field source). All other parameters besides the voltage have been kept constant, especially geometrical parameters.

Fig. 29, continuation: The overlay of all three traces is shown for the sake of illustration in the fourth picture, which is coloured. The overlay is possible because all videos have been done without any change of geometry. Each trace has its colour.
With the naked it is not observed, that the rotor is always moving laterally during its rotation. Only the measurement of the rotors path displays this fact together with irregularity of the paths, which is due to the self-adjustment mechanism.
Please look especially to the blue trace (at 13 kV), from which can be seen that the path changes from turn to turn remarkably, even it the voltage is kept constant.
The importance of the self-adjustment mechanism can be understood by comparing the extent of the x- and y- scale with the diameter of the rotor, which is only 64 mm.
The self-adjustment mechanism is necessary that the rotor can rotate at all !

Obviously, a permanent continuous adjustment of the lateral position of the rotor is indispensable during the rotation, otherwise the rotor would not spin. This experience was also reproduced with the large rotor of fig.9, which has a diameter of 46 cm. There the experience was made as following: The 46cm-rotor has an iron rod in its centre, marking its axis of rotation (thickness of the rod 2.2 cm). Swimming on the surface of water (with free lateral movement) it could be brought to rotation easily. But as soon as this free lateral movement was constrained by guiding the iron rod axis with a glass cylinder (such a cylinder with a diameter of 44 mm was put into the water below the surface so that the iron axis could not escape from it), the rotor was able to spin only as long as the iron rod did not touch the walls of the cylinder. When the iron rod touched the glass wall of the cylinder, the self-adjustment was stopped and with it the rotation. If the cylinder was shifted by hand for few centimetres, the self-adjustment could be continued and with it the rotation continued until the rod again touched the cylinder and all movement again stopped. This means that a circle of 44mm of diameter is not large enough for the self-adjustment mechanism necessary for the 46cm-rotor.

So it is seen, that a rotor with a diameter of 64 mm needs a lateral freedom of movement of about 6...7 mm and a rotor with a diameter of 46 cm needs a lateral freedom of movement of more than 44 mm – if the self-adjustment mechanism shall work, which is necessary that the electrostatic rotors under investigation in the present work can be driven by an electrostatic field. Because of this observation, it is necessary to ask for the reasons:

This could be inhomogeneities in the field and imperfections of the mechanical assembly.

The working mechanism which enables the rotation because of permanent continuous (self-) adjustment of the lateral rotor position is obviously the following: The rotor moves laterally until it reaches the position of the minimum of energy (potential) within the driving field, so that the Coulomb-forces are now free to produce a rotation in connection with the conversion of vacuum-energy. If this self-adjustment is not allowed (which is the case for an axis rigidly fixed at one position), the Coulomb-forces will also try to find a minimum of energy (potential), but they can only do this by producing a rotation (around the fixed axis), which stops as soon as the rotor blades are orientated along the gradient of the field.

The situation is like that:

- If this self-adjustment is working, the rotor runs into a global major minimum of the (electrostatic) potential being there ready for the conversion of vacuum-energy.
- If this self-adjustment is obstructed, the rotor runs into a secondary minor minimum of the (electrostatic) potential and is kept there not being ready for the conversion of vacuum-energy as long as it is kept there.

This perception is also confirmed by an experiment at which a rotor on a toe bearing (which fixes the axis of rotation) was mounted below a field source inside a vacuum chamber at a pressure of few $10^{-6}\,mbar$ [Bec 08/09] as shown in fig.30. A special feature of this vacuum recipient was a xy-manipulator, which allowed a lateral shift within the xy-plane that could be operated by hand (just by turning a hand wheel) in the time when the vacuum chamber was evacuated. Moving the rotor within the xy-plane in such manner gave the possibility to make the rotor spin for a fraction of a full turn (from few angular degrees up to three quarters of a full turn, but never a full turn or more), but it was not possible to obtain a permanent continuous rotation. After the experiment was finished, this became clear and the reason became understandable (the experiment of fig.29 was done later then the experiment of fig.30): The adjustment by hand is not able to produce a path as plotted in fig.29, so it was not possible to emulate the path of the self-adjustment mechanism by hand. It only would be possible, if we would know this path, but therefore a very special organized measurement would be necessary. When we move the rotor in xy-plane back and forth, the rotation goes back and forth, indicating, that the rotor blades are just kept within some secondary minor minimum of potential.

Fig. 30 shows the rotor, which was mounted inside the vacuum chamber and its attachment at a side flange of the vacuum chamber. During the operation of the rotor inside the vacuum, it is seen distinctly that the rotor wants to tilt. But the toe bearing was made with a rather long metal dome so that the rotor could not fall down. So the rotor tilted until the dome of the toe bearing touched the steel needle in the centre of the bearing. For the purpose of comparison, the diameter of the dome (beginning with 1.0 mm) was enlarged by drilling, but the principle remained unchanged: The rotor tilted as much as possible.

Fig. 30: Rotor on a toe bearing with a steel needle in the centre. The steel needle was isolated against the vacuum chamber with ceramic, so that the electrical connection of the steel needle was done with a thin copper filament.
The xy-manipulator in the photo is not yet this one, which allowed the xy-adjustment in-situ when the rotor was inside the vacuum. But in the further course of the experiment an xy-manipulator was mounted, which allowed the lateral movement of the rotor in the xy-plane with the vacuum chamber being evacuated.

Conclusion:

The principle of the conversion of vacuum-energy is operable, it is explained in Physics and it is verified – as described in section 4. This means that the fundamental Physics of the conversion of vacuum-energy is developed and concluded with the present work.

Technical applications and implementations of the principle developed with the present work require an optimization of the setup, especially of the mechanical precision of the rotor, the field source and their mounting together (relatively to each other) with the further development of the bearing. Also the homogeneity of the driving field is a requirement to future optimization. These suggestions for the future approach can be substantiated as following:

Obviously inhomogeneities of the field disturb the rotor with fixed axis at its continuous permanent rotation, because the field inhomogeneities keep the rotor at a position of a minimum of the potential – if there would not be any field inhomogeneities, there would not be any minimum of the potential, which could attract and hold the rotor. Consequently, it has a high priority to optimize the homogeneity of the driving (electric or magentic) field, if the rotor shall be operated with a fixed axis.

But in reality, the homogeneity of a field can never be ideally perfect, so the asymmetries of the rotor have to be minimized also. (But it is not possible to avoid asymmetries, because the working principle of the rotor needs asymmetries, for instance at the special assembly of the rotor blades with a special angle. It is only possible to avoid unwanted asymmetries.) The less unwanted asymmetries the rotor has, the less it will react to inhomogeneities of the field. But not only the shape of the rotor has to be optimized, but also its parallelism relatively to the field source. The plane of rotation has to be very exactly parallel to the field source, otherwise the rotor would be kept at a position at which one rotor blade finds its minimum distance to the field source. (This was also unproblematic for the swimming rotor, because it can only move always exactly on the flat surface of the liquid.)

A further disturbing factor comes from the metallic inner surfaces of the vacuum chamber, which are asymmetric by principle because of flanges, tubes, windows, etc... And these metallic inner surfaces also guide the flux lines of the electric field. Perhaps it is worth trying to shield the rotor from the inner surfaces of the vacuum chamber, but then the symmetry of the shielding has to be done with good perfection in order to avoid field asymmetries due to the shielding.

Finally, a further possible optimization can be the choice of the bearing of the rotor. Up to now the fixation of the rotor's axis was always done with a toe bearing (because it produces very low friction), which allows the rotor to oscillate laterally like a pendulum. Thus, the rotor can follow transverse forces to some degree, but not in the sense of the self-adjustment mechanism. With a double toe bearing or with a ball bearing this pendulum movement could be suppressed, but is this really a solution for the problem of the missing self-adjustment mechanism (at the fixed axis) ? Of course not, because it does not allow the rotor to find the global minimum of the potential by lateral movement.

The possible amplitudes of the lateral pendulum movement can be seen if they are not suppressed. This was additionally analyzed with a rotor hanging on the tip of a needle which was held by a magnet (by magnetic attraction) and which was driven by electrostatic forces. (The field source was standing on the table below the hanging rotor.) As expected from fig.29 the rotor rotates (when the voltage is switched on) and it also performs a very strong tumbling motion, which is extremely irregular and not slow. (It can be asked whether this might be a chaotic motion.)

As soon as the disturbing transverse forces are smaller then the driving forces from the conversion of vacuum-energy, the rotor will spin continuously and permanently. Future will show, whether the suggestions given above will already be enough to reach this goal. Despite all technical difficulties with the fixed axis of rotation, it should never be forgotten, that the principle for the conversion of vacuum-energy is invented and verified in Physics. It is clear that it works. And if mankind could not bring it forward to the fixed axis, the reason would not be the principle and not Physics but technical unhandiness.

5.3. Investigations by others

machinery to The fact that measurable and useable energy can be extracted from the zero-point oscillations of quantum theory (which can be converted to a classical type of energy) is not yet accepted by everybody. However this lack of acceptation is somehow astonishing, because the extraction of energy from the zero-point oscillations is already theoretically understood and experimentally confirmed (as mentioned above) with the Casimir-effect to an accuracy of 5% in the measurements [Lam 97].

Furthermore the existence of some "dark energy" is well-known from its influence on the expansion of the universe because it is equivalent to a mass causing gravitation. It is now under discussion, whether this "dark energy" in the universe might perhaps be the same as the "energy of the zero-point oscillations" of quantum theory. In order to test, whether there is a connection between these both types of energy, [Bec 06] made a proposal which initiated an experimental investigation being under operation just now [War 09]. But independently of the fact whether the "dark energy" of the universe can be identified with "the energy of the zero-point oscillations" or not, it is well known from the Casimir-effect, that the mere space (this is the void) definitely contains the energy of the zero-point oscillations. The logical consequence is that the zero-point energy is at least one contribution to the "dark energy" of the universe.

Now, since the energy of the zero-point oscillations is recognized, it should be found, how we can get benefit of it. First of all, it is interesting how to convert this energy into a classical form of energy, which can be used for energy supply.

Two really nice books with an overview over the use of the zero-point energy should be mentioned. They are written by Tom Valone, and they reflect the actual state of development world-wide: [Val 08] is popular scientific book, whereas [Val 05] is made for scientists. Some of Valone's references lead to some of the work reported on the following pages.

Let us come now to a description of several theoretical and experimental investigation with regard to the conversion of the zero-point energy of the vacuum or some other unknown energy like the dark energy of the universe. We will follow roughly a chronological order.

• <u>Nikola Tesla</u>

The first who recognized the energy of the mere space and spoke about its utilization was the great Nikola Tesla. Already in 1891 he addressed the words to the American Institute of Electrical engineering:

"Ere many generations pass, our machinery will be driven by a power obtainable at any point in the universe. This idea is not novel... We find it in the delightful myth of Antheus, who derives power from the earth; we find it among the subtle speculations of one of your splendid mathematicians... Throughout space there is energy. Is this energy static or kinetic.? If static our hopes are in vain; if kinetic - and this we know it is, for certain – then it is a mere question of time when men will succeed in attaching their the very wheelwork of nature."

Today it is almost impossible to reproduce this part of Nikola Tesla's work, which is concerned with the zero-point energy, at least because there are hardly any reliable publications available. It is said, that Tesla was able to transport energy trough the space of trough the surface of the earth (or trough the atmosphere ?). He got some patents for this type of energy distribution [Hee 00]. Furthermore it is known, that Tesla built up the „Wardenclyffe Tower" on Long Island in 1903, from which is said, that he could distribute energy without any cables even on the long distance to Europe. It is said furthermore, that J. P. Morgan stopped this development of energy distribution, because it was not clear how to put a meter on it to make an electricity bill [Val 08]. To what extent Tesla can be regarded as a pioneer of zero-point energy is not clear. There is an anecdote telling that Tesla modified a car ("Pierce Arrow") in a way that he could drive without gasoline. But the car disappeared together with the "box" supplying the car with energy. And the only publications about this project is said to be the daily press of Buffalo in 1931. Of course there are no scientific articles there. The reason, why Tesla is mentioned here is, that he expressed his expectation, that it should become possible to utilize the energy of the void.

• <u>Casimir and work relating directly to him</u>

It is well known, that Hendrik Brugt Gerhard Casimir in 1948 postulated the force which was given his name later. The basis of this force are the zero-point oscillations of quantum theory which occur also with electromagnetic waves. According to his theoretically findings, a force between two parallel metallic conducting plates should occur even if they are electrically uncharged [Cas 48]. Only for very small distance between the plates, those forces should be noticeable. Thus, the expectation was that the plates have to be arranged in a distance in the sub-micrometer range, otherwise the Casimir-effect could not be experimentally verified. At the time of Casimir, the experimental challenge was immense, to mount two macroscopic plates that close to each other with adequate precision, and to measure the very small force between them. Consequently, Casimir's considerations had not been taken serious for a rather long time. Also measurements performed independently by two experimental groups in 1956 and 1958 could not lead to acceptance for Casimir's work, because the precision of the measurement was not good enough. One group was in Russia [Der 56] after theoretical preparation by [Lif 56], and the other group was [Spa 58] (whereas Spaarnay came to an uncertainty of measurement of about 100%). Only in 1997, nearly half a century after Casimir's publication, and experiment done by Steve K. Lamoreux [Lam 97] came to a precision of about 5%. This had the consequence, that the Casimir-effect became accepted from there on. Since that time it is generally accepted that the zero-point oscillations of electromagnetic waves cause measurable forces, which we can feel in our macroscopic world.

In the very recent days, due to the upcoming Nanotechnology, the Casimir-effect begins to be of technological importance, because the Casimir-forces increase with the reciprocal of fourth power of the distance between the surfaces ($F = \frac{A \cdot hc\pi}{480 d^4}$, with $F = \text{force}$, $A = \text{area of the casimir-plates}$, $d = \text{distance of the casimir-plates}$). Thus the Casimir-forces can dominate on very short distance [Lam 05]. Actually modern Nanotechnology more and more faces the problem, that surfaces coming too close to each other of touching each other are attracted so much, that they can be separated only destructive.

Such a clear confirmation of the energy of the zero-point oscillations arise hope, that we can find a way to use it not for destruction but for a production of energy without polluting our environment.

And here is an explanatory note for those who doubt:

Sometimes there is a counter-argument against the possibility to utilize the energy of the zero-point oscillations practically: The zero-point oscillations are the ground-state of energy, this is the lowest possible state. Thus it is impossible to go down to an even lower state, because there is no lower state. Because we can not get energy out of the ground-state, it is impossible to extract energy from the zero-point oscillations of the electromagnetic waves of the vacuum.

This argument is wrong. The reasoning is the following:

- If the argument would be correct, Casimir also could not get energy out of the ground-state. But actually his effect does.
- But why can the Casimir-effect extract energy out of the ground-state? Because some of the waves of the electromagnetic zero-point oscillations are switched off.
- The question is of course: Do we have several more possibilities to extract energy out of the ground-state of the zero-point oscillations ?
- The answer is rather simple: Energy can be extracted if the waves of the zero-point oscillations are switched off or if they are altered somehow.
- The above counter-argument would be sensible only, if it would be impossible to switch off or to alter the waves of the zero-point oscillations, as we know it from typically quantum mechanical systems (like the energy-spectra of the electrons in the atomic shell). But this is not the case for the zero-point oscillations of electromagnetic waves.

And this is beeing realized: The waves are switched off or altered:

- Casimir (and Lamoreux and many others) switch off some of the zero-point oscillations and convert their energy into classical mechanical energy.
- In the work presented here, the wavelengths of the zero-point oscillations are changed (by applying an electrostatic or a magnetostatic field), and the energy coming out by changing the wavelengths is converted into classical mechanical energy.
- Perhaps physicists will find several other ways how to manipulate the zero-point oscillations (not only the wavelength) in order to gain energy from them ?

• Stochastic Electrodynamics (SED)

The basis is a philosophical attempt, namely the idea to explain well-known and proven quantum-mechanical effects without the application of the usual principles and formulas of quantum theory. Even if is often regarded as an "exotic approach", its results fit into accepted physics without any contradictions. The considerations initiating Stochastic Electrodynamics originate from Trevor W. Marshall and T. Brafford in the early 1960ies, have been further conducted by Luis de la Pena and Ana Maria Cetto [Col 96], [Pen 96] and undergo a first bloom with Timothy H. Boyer [Boy 66..08]. A review is also given by Boyer in [Boy 80] and [Boy 85].

A second blossom comes to Stochastic Electrodynamics due to a group of Calphysics Institute (Bernard Haisch, Alfonso Rueda, Harold E. Puthoff, Daniel C. Cole, Michael Ibision, only to mention some of these scientists), when they begin to think about several very fundamental question in Physics (as for instance about the origin of the mass and inertia) and include Stochastic Electrodynamics into their considerations. Finally their work implies possible applications, such as utilization of the energy of the zero-point oscillations or deep space travel [Cal 84..06].

The approach of Stochastic Electrodynamics can be described roughly as following:

Basic assumption is the postulate that the zero-point oscillations (which actually result from quantum theory) exist, and that their spectrum is the same as we know it from quantum theory. Further assumptions of quantum theory are not needed. If now the behaviour of this zero-point radiation and its interaction with all visible matter is regarded, of course matter absorbs energy from the radiation of the zero-point oscillations during all of the time of its existence and emits the same type of radiation at the same time, because each elementary particle does its zero-point oscillations. (By the way: This idea is very similar to the postulate of my work, that charged elementary particles (as sources of electric field) are supplied with energy from the vacuum.)

On the basis of this approach, all phenomena known from quantum theory can be explained complete without the use and without the principles of quantum theory !

The black body radiation with its characteristic spectrum as a function of temperature results from the movement due to the zero-point oscillations of the elementary particles forming the body. Also the photoeffect is explained within Stochastic Electrodynamics. In the history of quantum mechanics, one of the prominent results was the explanation of the energy levels of the electrons in atomic shell. In the formalism of Stochastic Electrodynamics, stable states (at which electrons can stay) are achieved when the energy being emitted from the electrons because of their circulation around the nucleus, is identically compensated by the energy which they absorb from the zero-point radiation of the vacuum. (This contains an explanation, why the electrons do not fall into the nucleus because they lose energy due to their circulation. There is some analogy with Bohr's first and third postulate, according to which stable states of shell-electrons are only possible for constructive interference of the electron-waves.) And finally it should be said, that the equilibrium between absorbed and emitted radiation (in Stochastic Electrodynamics) leads to the same discrete energy-levels as we know them from quantum mechanics.

Not only the results of quantum mechanics but also the results of quantum electrodynamics are reproduced with the formalism of Stochastic Electrodynamics, for instance such as the Casimir-effect, van der Waals – forces, the uncertainty principle (which has been derived the first time by Heisenberg) and many others.

For the sake of completeness it should be remarked, that Stochastic Electrodynamics of course explains the phenomena of nature on its own, not trying to reproduce the mathematical structure of quantum theory and even not in connection with the formalism of quantum theory. So for example the famous Schrödinger-equation, as a typical formula of quantum theory can not be derived with the means of Stochastic Electrodynamics, because such a formula simple is not a topic of Stochastic Electrodynamics. In this sense Stochastic Electrodynamics and Quantum Theory are two independently concepts, which describe the same phenomena of nature, but which have totally different philosophical background.

Of course Stochastic Electrodynamics is not as widespread as Quantum Theory. But it is in complete confirmation with all nowadays known phenomena of nature. Thus it is sensible to accept it for further considerations of "how to extract energy from the zero-point oscillations", which can lead to interesting results, because new thoughts might emerge. Buy the way: The wrong counter-argument of ground-state energy quoted above is not a problem for Stochastic Electrodynamics, because it is obvious, that the zero-point radiation can be altered.

By the way: Not only the work presented here proves, that it is possible to extract some energy from the zero-point oscillations of the quantum vacuum. Also in [Col 93] Daniel C. Cole and Harold E. Puthoff come theoretically to the result, that this type of energy can be utilized. A suggestion how to gain this energy is shown in the US-patent [Hai 08].

• Vacuum-fluctuation Battery by Robert L. Forward

Already in 1984 Robert L. Forward developed his idea of a "vacuum-fluctuation battery" [For 84]. According to his suggestion, this would be a spiral coil made of a conducting foil (see Fig. 31) which is charged electrically. If the spiral is small enough (e.g. with micromechanical manufacturing) that the individual convolutions come close to each other in the submicrometer range, two forces will arise: A repulsive Coulomb-force (due to the electrostatic charge) and an attractive Casimir-force (due to the close distance of the spiral convolutions). It would be possible now to apply exactly such amount of electrical charge, that the Casimir-forces and the Coulomb-forces hold themselves exactly in balance.

If on the contrary slightly less electrostatic charge is applied, the spiral will be compressed by the Casimir-forces, so that the energy of the electrostatic field increases due to Casimir-force. This additional field-energy (of the electrostatic field) is extracted from the zero-point oscillations of the vacuum.

But in the opposite, if some additional electrostatic charge is applied, the spiral will expand. This is the procedure, which Robert Forward calls "recharging the battery". An oscillation of the spiral length would lead to an alternating current. Robert Forward now poses the question, whether the Casimir-forces are conservative or not [For 96], which has a direct influence on the question, whether on the basis of his suggestion zero-point energy can be converted into electrical energy or not. It looks like the Casimir-forces are not conservative.

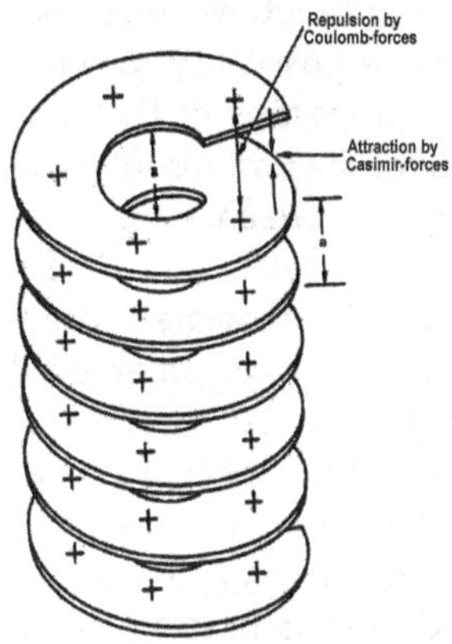

Fig. 31: Screw-shaped spiral made of conducting foil and being electrically charged. It is under discussion as a possible setup to extract utilizable energy from the alternating field of the zero-point oscillations of the vacuum (see [For 84]).

As a consequence of his considerations, Robert L. Forward points out the possibility the use the zero-point energy of the vacuum as an energy source. But the setup requires micromechanical manufacturing technology. Thus it could not be analyzed experimentally up to now. Nevertheless his idea to support the Casimireffect by electrical means resembles to the method introduced in my work (which does not require micromechanics).

• Manipulation of the Casimir-forces and Casimir-Energy

Several recent investigations demonstrate that Casimir-forces can be manipulated by suitable techniques (e.g. like different materials for the surfaces, differently shaped surfaces, suitable cavities, suitable distances, etc....). The manipulation can be so drastically, that it is even possible to switch between attractive and repulsive forces (see for example [Mun 09]). Fabrizio Pinto has got an US-patent regarding the modulation of the strength and the direction of Casimir-forces between two surfaces [Pin 03]. Therefore he uses two Casimir-plates. For one plate the dielectric properties are most important, for the other plate its magnetic properties are most important with regards to Pinto's construction. Now the dielectric and the magnetic material properties can be influenced for instance with a laser (via Kerr effect, via Pockels-effect for dielectric properties and via Faraday-effect, via Cotton-Mouton-effect for the magnetic properties, and so on...). This gives a possibility to alter the Casimir-forces. The influence of the material properties of the surfaces can be so extensive, that not only the strength of the Casimir-force can be altered, but even its direction can be changed between attraction and repulsion. Pinto's theoretical computations as well as his thought-experiments come to the result, that it should be possible to develop closed paths along which the total work is not zero. With other words: It should be possible to arrange a Casimir-setup with Casimir-forces in a way, that a non-conservative cycle is possible. And this cycle, according to Pinto can be arranged in a way, that it produces classical energy (coming from the zero-point oscillations of the vacuum) [Pin 99]. As soon as it would be possible to build a machine, which allows passing this cycle endlessly, a continuous conversion of zero-point energy into classical (mechanical) energy would be possible.

Similar results can also be seen from Werner Sprenzel and Sven Mielordt in 1985 [Spr 85], but at this time they did not have exact quantitative computations like Pinto has.

For a concrete realization of a Casimir-engine, Pinto suggests micro-mechanical manufacturing because of the very small dimensions in the range of 50... 100 µm. The Casimir-plates would have to oscillate with a frequency in the range of about 10kHz. The energy-density coming from his suggested machine should be expected to be in the range of about 10µJoule/cm^2, corresponding to net power of half a nanoWatt with plates of about 1 cm^2.

It should be mentioned that also [Cou 99] theoretically expects the possibility to manipulate Casimir-forces by the use of magnetic fields.

It might be that Casimir-forces could be really controllable. Experiments can be expected with interest.

• Energy from quantum-noise

Noise contains energy. Electrotechnicians use it, when they build noise-generators. For instance thermal noise of a pn-junction can be amplified to an arbitrary strength. Of course it is no problem, to extract energy from such thermal noise. This leads to the question, whether noise stimulated by the zero-point oscillations of quantum theory can be a used for the extraction of energy as well, perhaps after being rectified.

If one wants to prove the extraction of energy from the quantum-noise, thermal noise must be suppressed. Because the intensity of quantum-noise is not very high, this has the consequence, that it's measurement requires cooling to the temperature of absolute zero (T=0K). The noise remaining at this temperature is for sure not thermal noise. At T=0K, the only noise is quantum noise (for instance due to zero-point oscillations).

Thus the task is well defined:

A possibility has to be found in order to verify the movement of charge carriers in the direct proximity of the temperature of absolute zero. Already at the early 1980ies, Koch et. al. analyzed the quantum noise in Josephson junctions at first theoretically [Koc 80] and then experimentally [Koc 82]. It is not yet clear, whether this noise originates from the zero-point oscillations of electromagnetic waves in the vacuum. Christian Beck suggests an experiment which he hopes to clarify this question ([Bec 05] and [Bec 06]). If a connection between the noise in Josephson junctions and the zero-point quantum-noise would turn out be existing, consideation about a possibility to extract energy from it will be necessary. The problem is the very low energy-density of the zero-point quantum-noise. Tom Valone thinks about the possibility to investigate, whether it might be possible to rectify such quantum-noise with some very special diodes in order to get an electric current out of it [Val 08].

• Rotating spheres by Wistrom and Khachatourian

Certain parallels can be drawn between their work and the vacuum-energy rotor presented in the publication here, even though Wistrom and Khachatourian do not yet have theoretical explanations for their observations and although they do not yet have quantitative measurements of their machine-power.

Already in 2001 (published in 2002) Wistrom and Khachatourian of the University of California ([Wis 01], [Wis 02]) found forces between electrostatically charged spheres, which can not be explained the classical theory of Maxwell's electrodynamics. Similar like in my work a supplement to Coulomb's law is necessary to explain the rotation of the rotor, Wistrom and Khachatourian need a supplement to Coulomb's law to explain the rotation of their spheres. Details: My supplement necessary for explanation of the vacuum-energy rotor is the Theory of Relativity, supposing that the speed of light is an upper limit to every speed at all, and from there concluding that also electrostatic fields (and magnetostatic fields the same) propagate with finite speed, not faster than the speed of light. The supplement to Coulomb's law, which Wistrom and Khachatourian need to calculate the forces and torques in their experiment are based on the Gaussian law, which they apply to the electrostatic potential. On this basis they are able develop a proportionality-equation for the torque, coming to the result quoted in (1.70). This can be seen in analogy to equation (1.53) ... (1.60) stated above.

An illustration of their setup is shown in Fig.32, which is a reproduction of a drawing in their original publication [Wis 01].

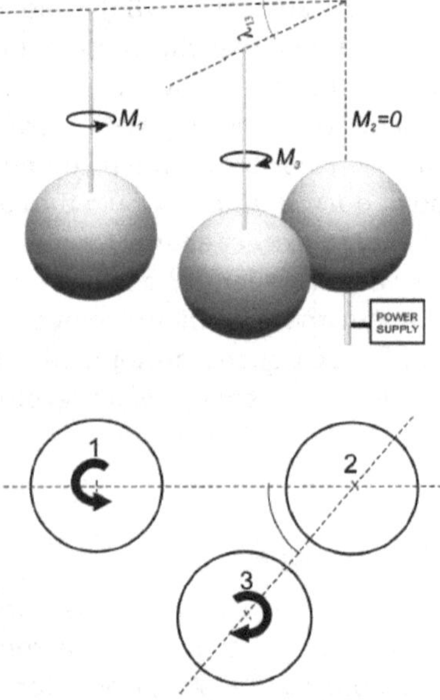

Fig. 32: Setup consisting of three electrostatically charged spheres, which take up a torque only because of the charge on their surfaces. The spheres have diameters of 27 cm. The voltage applied was varied in the range of 400 ... 5000 Volts. The wire-filaments on which the spheres are hanging have diameters of 127 µm.
The image is taken from the original publication of [Wis 01].

What Wistrom and Khachatourian still have to find is an explanation for the origin of the energy driving the rotation of their spheres. They say, that any electric current is negligible, and they furthermore state that the torque is caused by electrostatic reasons. They also say that there are no magnetic fields acting onto the spheres. The spheres only perform rotations but no translations. The observed torque makes two of the three spheres spin until finally the wires on which those spheres are hanging are twisted so much, that the torque caused by the torsion of the wires do stop the rotation of the spheres. It should be mentioned that the torque acting on the electrostatically charged spheres remain there over many hours and days, which was tested for a period of time of even 200 hours. This demonstrates that there is some energy driving the spheres, which does not exhaust the electric power of the electric charge brought to the surfaces of the spheres.

The theoretical determination of the torque as stated above, and given in (1.70), is in agreement with the experimental observation, which can be understood as an additional confirmation of the fact that the forces really have electrostatic nature. Unfortunately Wistrom and Khachatourian up to now did not do a quantitative measurement of the machine power for the comparision of the electric power and the mechanical power. But from the fact, that the electrostatic charge is not lost during more than a full week of time, it can be concluded, that the produced mechanical power is more then electrical power loss (due to imperfections of the isolation).

$$M \propto \frac{1}{4\pi\varepsilon_0} \cdot \frac{V^2}{a^5 \cdot h^4} \quad \text{with} \quad \begin{aligned} V &= \text{potential} \\ a &= \text{radius of the spheres} \\ h &= \text{distance of the spheres to each other} \end{aligned} \qquad (1.70)$$

By the way: [Hen 04] checked the theoretical computation of the Coulomb-forces and the torque given in (1.70), but they could not confirm the result of Wistrom und Khachatourian. Nevertheless they did not contradict the experimental result of Wistrom und Khachatourian, as stated by [Taj 08]. An upcoming discussion (by [Hen 04]) about possible asymmetries as a reason for the rotation of the spheres is unfounded, because the spheres perform many hundreds of full rotations before they are stopped by the torque caused by the wires, as it was reported by personal observation of the setup in operation by [Rei 09]. Asymmetries in the setup can only cause a rotation of less than one full turn (i.e. less than 360°) by principle.

As a survey it can be said that the observations by Wistrom and Khachatourian confirm the results of the work presented here, although their theoretical considerations are not yet complete and although the confirmation of the conversion of vacuum-energy by quantitative measurements of the electrical power and the produced machine power is still to be done.

• Vacuum-rotor by NASA and Gravitec. Inc.

Among all the work referred to in section 5.3., this is the investigation most close to mine.

There was a cooperation between the American space agency NASA and Gravitec Inc in the year 2003, peforming tests inside the vacuum chamber NSSTC LEEIF at the Marshall Space Flight Center. Due to confidential reasons, the results have been published only in December 2007 [Nau 07]. The setup was an asymmetric capacitor, which was built in a way that one plate of the capacitor was fixed at an axis mounted on a rotating arm with a torsion wire used as a rotational axis. The torsion wire with the asymmetric capacitor plate was connected with a high voltage supply, which was variable from 0 kV to 45 kV. The voltage could be applied under atmospheric pressure as well as under a vacuum of $1.72 \cdot 10^{-6}\,Torr$. It was applied between the torsion wire with the capacitor plate on the one hand and the all the other parts of the vacuum chamber on the other hand.

The observed results have been the following:

At atmospheric pressure (i.e. without vacuum) there is a torque bringing the capacitor plate into rotation until finally the torque of the torsion wire stops the rotation. Switching the electrostatic high voltage on and off in appropriate moments makes the capacitor plate fulfil a pendulum movement of rotation, but unfortunately the video (done as a part of the record of the experimental documentation) never displays one full turn of rotation or more than one turn of rotation. Furthermore [Nau 07] states that the torque causing the rotation of the capacitor plate directly increases with the high voltage and also this torque completely disappears at a voltage of zero.

At a vacuum pressure of $1.72 \cdot 10^{-6}\,Torr$, the same type of pendulum movement of rotation as observed under air can be seen, but the torque is much smaller than under atmospheric pressure. According to the knowledge our equation (1.68) this is not surprising.

Regrettably the NASA-work did not display the continuous rotation of the capacitor plate over several revolutions, and furthermore there is a lack of quantitative measurements of the electrical power and the produced machine power, as it would be necessary to verify the interaction of the capacitor plate with the inner structure of the vacuum, respectively to verify the conversion of vacuum-energy. Nevertheless the NASA-work confirms the results shown in the work presented here.

• Arkadii A. Popov

Popov derived theoretically (considering gravitation) self-consistent solutions of Einstein's field equations, that vacuum-fluctuations can create Lorentz'ian wormholes. Therefore he took interactions between vacuum-fluctuations and vacuum-polarization on the one hand and the curvature of spacetime on the other hand into account, which is the well-known gravitation [Pop 00], [Pop 05]. In a theoretical work [Liu 93] demonstrates, vacuum-fluctuations can create Lorentz'ian wormholes.

On the basis of the awareness that even gravitation interacts with vacuum-fluctuations it is not surprising any further, that electrostatic and magnetostatic fields do the same (as used as a basis for the possibility to convert energy from the mere space of the vacuum as presented here).

• Energy-converter by Hans Coler

If the literature-search is expanded from the "zero-point energy of the vacuum" to all types of "energy-source unknown up to now", a lot of work can be found, which does not allow proper evaluation. In order to avoid confusion, it would be simplest to ignore this type of work at all. However then the question would remain openly, whether something useful might be lost.

As an example, one of the layman's trials shall be mentioned here, namely that of Mr. Coler, which neither applies scientific terminology nor uses the usual documentation (as often observed in such type of work). The situation somehow reminds to exploration of the earth by sailors and geographers in the 16th and 17th century. They found new, previously undiscovered land (analogous to our quest for new energy sources not previously known), and they asked the natives, and they heard stories far from the scientific terminology. Nevertheless they followed the tracks they got carefully, until they finally came to useful results.

About the Coler-invention, such a report is available as a military intelligence report from the Second World War, which today due to the elapsed time is no longer inaccessible [Hur 40]. Therefrom a report can be seen nowadays by [Nie 83] and [Mie 84], according to which Hans Coler (partly in collaboration with Mr. von Unruh) had built two types of machines in the years from 1923 to the 1930s. He called one those both machines "Stromerzeuger" and the other on "Magnetstromapparat" (literally translated "electrical current generator" and "magnetical current apparatus"). Both type of machines are said to have an efficiency of more than 100%, whereas the "electrical current generator" required a power-supply but the "magnetical current apparatus" did not require and supply at all. Coler brought his machines to several specialists, among them scientists at the Universities of Berlin, Munich, Trondheim and Copenhagen. He performed his demonstrations and the scientists did their own analysis. Finally the functionality of his devices have been seriously confirmed for many times. The problem is that nobody could explain the reason of the operation (the operating principle). A further problem is that Mr. Coler's terminology was totally from the scientific one, so it was not understandable for professionals. Consequently the reports referred to in [Hur 40] appears rather confused and has no valuable explanatory power to us. Anyhow, it is clear, that the generated electrical power is more than the input power of the machines. The very first machines of the Coler-type produced an electrical power in the range of few Watts, but during the years Coler was able to build larger machines with a power-output in the range of several KiloWatts.

A description of both Coler-machines can be found today in the most convenient at [Wik 09], but also there the functional principles (as mentioned above) can be not be explained in our terminology. For this reason, the Coler-machines shall not be discussed in detail in the present work here, because a scientific presentation of his results is not possible nowadays. A popular scientific approach to Coler's machine can be found in [Hec 09]. Strangely, no efforts were made to reproduce the work of Hans Coler nowadays. One attempt of reproduction is known from George Hathaway, who presented a replica of the "magnetical current apparatus" in 1981 on the "First International Symposium on Non-Conventional Energy Technology," producing a voltage of 50 mV. Supposedly an American group might have tried further reproductions of the Coler-machine, and it is said that they got 7.2 kW of electrical power, but there is hardly any information about this work to be found.

Despite all the ambiguities and unanswered questions, however, Coler's work demonstrates, that there is some hitherto unknown energy, which can be converted into a well-known classical energy - in the case of Mr. Coler into electrical energy. Whether this is the zero-point energy of quantum theory or some other type of energy can not be clarified in the unclear situation now. In any case, it seems desirable to reproduce the work of Mr. Coler in order get further and detailed investigations about them, which hopefully might lead to an understanding in our terminology.

Further discussion of "every work which can be found" regarding "energy-source unknown up to now" is not helpful here, because it is not possible to get an understanding in our usual terminology from it.

5.4. Outlook to imaginable applications

If the present work should only be of importance for fundamental Physics, namely for the verification of the theoretical models described in the sections 2 and 3, then the successful experiments of section 4 would already be the final execution of all tasks. In reality, the possible expected benefit of the work now begins:

Rotors for the conversion of vacuum-energy are driven by static fields, so they do not consume classical energy. They convert a type of energy, which was not respected very much up to now (which can be called vacuum-energy) into classical mechanical energy (mechanical rotation). This arises the hope for an application as an inexhaustible source of energy, which moreover has the advantage not to pollute our environment at all. How inexhaustible this source of energy really is, can be seen from the modern standard model of cosmology [TEG 02] [RIE 98], [EFS 02], [TON 03], [and many others], among whose statements we find also the consistency of our universe. According to this standard model the universe consists of

- 5 % well known particles, this is all matter, which mankind can see,
- 30 % invisible matter, these are elementary particles unknown up to now,
- 65 % vacuum-energy.

So mankind could have the largest dominant part of the universe as a source of energy – if we learn how to get it. Of course not everybody can imagine that this source of energy begins to be available for mankind (for instance [Bru 06]), but there is also very much substantiated work and there are many serious colleagues, who come to the result, that the conversion of vacuum-energy is indeed possible (for instance [Sol 03], [Sol 05], [Sol 06], but also [Kho 08], [Red 08], [Kho 07a], [Kho 07b], [Put 08], [Ole 99]).

A practical and profitable benefit can be achieved from vacuum-energy (of the universe) from the moment, at which vacuum-energy conversion will deliver more energy than its operation consumes. As soon as this condition is fulfilled, vacuum-energy rotors can be of essential benefit for power supply industry as well as for protection of the environment. We want to make some few considerations about the "economical and environmental benefit on industrial scale" now, beginning with a comparison of the two types of rotors presented in the present work:

▪ **In the case of the magnetostatic vacuum-energy rotor**

this benefit presumes (after the endless rotation will work with a fixed axis), that the cooling of the superconducting rotor blades will consume less energy than the rotation inside a magnetic field of a permanent magnet produces. But the cooling of the superconductors requires liquid nitrogen, whose production needs energy. "Economical and environmental benefit" requires very good materials for thermal isolation, i.e. very good cryostats in order to minimize the amount of necessary liquid nitrogen.

▪ **In the case of the electrostatic vacuum-energy rotor**

the main expenses of energy to drive the rotor are for sure not the losses of electrical energy to maintain the electric field and to keep it constant. This is already clear from the very simple setup in section 4.4. Things are going rather like that: "Economical and environmental benefit" requires very good vacuum chambers and energy saving vacuum pumps.

By the way, an interesting alternative for the field source could come under discussion: An electret. This is a material, which can permanently produce an electrostatic field similar to a permanent magnet which permanently produces a magnetic field. Same as the permanent magnet has to be polarized magnetically before producing a magnetic field, the electret has to be polarized electrically before producing an electric field. If it would be possible to find an electret which keeps its polarization for a long time, it could be interesting to investigate, whether such a material can drive an electrostatic rotor for the conversion of vacuum-energy with its field. The production of a suitable field strength should be imaginable, because there are electrets available which can produce many Kilovolts per centimetre.

Available electrets can be found rather often among plastics [Wik 08], [Mel 04], as for instance Teflon (=Polytetrafluorethylen), Polypropylen, Polyethylenterephthalat, Polytetrafluorethylenpropylen, Polypropylen, Polyethylenterephthalat (PET-foil), Polyvinylidenfluorid, but there are also remanent dielectrics available such as for instance Siliziumdioxid or

Siliziumnitrid. Some of the mentioned materials have been tested in the present work. They have been electrically charged by rubbing their surface, and then they have been held over the rotor of fig.14 as a field source. The tests have not been quantitative. In some cases rotations have been observed, which normally starts rather speedy (this means with a large torque indicating a large field strength) and also ended rather abruptly. In the most cases, the angle of rotation was less than one full turn (less than 360°). This is a typical indication for an inhomogeneous charge distribution on the surface of the electret, which has the consequence that the rotor only finds the position of the minimum potential and comes to a standstill there. In very few cases rotations up to 2...3 turn have been observed [e9] (angle of rotation not being not reproducible). This might happen if the charge distribution on the surface of the electret is not too inhomogeneous. The problem is, that the electret is not a conductor and so the electrical charge can not distribute homogeneously on the surface by alone. We see that there are still many questions open before it will be clear, whether an electret can be used to drive a rotor for the conversion of vacuum-energy.

By the way: The electret which allowed a bit more than two full turns was an air-balloon made of an elastomer. It has been charged up by rubbing the surface with artificial leather so much, that electrical breakthrough occurred during the procedure of rubbing. The electrical charge was enough to keep the balloon at the ceiling of a room against gravitation for many hours (which can also be understood with the image-charge method explaining the attractive forces between the balloon (+q) and the ceiling (–q)). This also means, that the charge is not leaving the surface of the balloon. The rotation was reproducible but not the measurement of the angle. This should be due to the inhomogeneous charge distribution on the surface of the balloon, which also had the effect, that the balloon's capability to stick to the ceiling depends on the orientation of the balloon.

If the electret is regarded as a remanent polarized dielectric (in analogy with the permanent magnet, which is a remanent polarized ferroelectric), the electret does not produce an electric current, it does not ionize gas molecules of the air (as soon as the breakthrough is over). This perception is confirmed by the fact that the balloon can stick to ceiling for many hours keeping its electrical charge. This demonstrates the quality of the isolation of the material. An estimation of the energy balance and energy sum is given in [e9] coming to the result that the electrical charge which might flow away from the balloon does not support the rotor with enough electrical power to explain its rotation, already confirming the "over-unity" criterion for the first time. But at that stage of the development, quantitative measurements had not yet been performed with a precision as shown in section 4, thus this result is just mentioned casually.

Possibilities to enhance the mechanical power with regard to technical applications:

Let us start with the question: How much engine power has been produced with the vacuum-energy-rotors up to now?

▪ For the $46cm$-rotor swimming on water (fig.12) the mechanical power was analyzed to be $P = 1.75 \cdot 10^{-7} Watt$. From (1.1) it is known, that part of this power might be due to the recoil of gas ions, but the order of magnitude is clear: $P = 10^{-7} Watt$ have been converted from vacuum-energy. The voltage was in the range of $4 \ldots 7 kV$. Further enhancement of the voltage speeds up the rotation and thus enhances the mechanical power.

▪ For the $51mm$-rotor in the vacuum (fig.19) an estimation of the power was done on the basis of the fact that the attractive Coulomb-force lifts the rotor by about $h = 2 \ldots 3mm$ when the voltage is switched on. The rotor sinks back by the same height, when the voltage is switched off. This can be observed under air in the same manner as in the vacuum. The mass of this rotor is $m = 2.02\ Gramm$, thus the mechanical work for lifting the rotor is about $W = 40 \ldots 60\,\mu Joule$. The process takes roughly about $\Delta t = 1 \ldots 2 sec.$, so the related mechanical power is at least $P = W / \Delta t \geq 40\,\mu Joule / 2 sec. = 20\,\mu Watt$ (plus the power to surmount the friction due to the viscosity of the oil). The voltage for this test had been $16kV$.

▪ The most important rotor (diameter of 64 mm) for the exact power measurement is analyzed in section 4.4. It produced a power of $(1.5 \pm 0.5) \cdot 10^{-7} Watt$ at a voltage of $U = 29.7\ kV$. This is the power value with the most exact measurement, so it can be used for further considerations.

▪ An extrapolation to industrial dimensions (for "economical and environmental applications with benefit") can be done only roughly in orders of magnitude. More precise values are not available in the moment. Therefore, we can use the proportionalities $P \propto U^2$ and $P \propto R^2$ as shown above:

If we assume a voltage at an order of magnitude of $U \sim 10^4 Volt$ as a realistic presumption and a rotor radius at an order of magnitude of $R \sim 10^{-1} m$, the power at an order of magnitude of $P \approx 10^{-6}\,Watt$ leads to the following imagination:

Energy-converting rotor	R	U	$P \propto U^2 \cdot R^2$
Realistic presumption (as an average over the estimations shown before)	$10^{-1}\,m$	$10^{4}\,V$	$10^{-6}\,W$
Vacuum allows large breakthrough field strength (with appropriate distance from rotor to field source)	$10^{-1}\,m$	$10^{7}\,V$	$10^{0}\,W$
Large rotors could be assembled (20 meters of diameter inside a building)	$10^{+1}\,m$	$10^{7}\,V$	$10^{+4}\,W$
Several rotors could be piled up in a cascade (10 rotors one upon the other inside a building)	$10^{+1}\,m$	$10^{7}\,V$	$10^{+5}\,W$
This could lead to dimensions of "economical and environmental benefit on industrial scale".			

6. Summary

If the Theory of Relativity is taken serious, we have to accept that electrostatic same as magnetostatic DC-fields propagate not faster then with the speed of light, thus with finite speed. This means that the field of every field source together with its energy (of an electric charge of an electret in the way same as the field of a magnet) is not everywhere in the space at the same moment, but it fills the space beginning from its source. This conception does not depend on the question at which moment the field source (for instance the electric charge) has been "born" (which might have been happened at the "big bang"). Thus we see a flux energy coming out of the field source.

If this flux of field-energy into the space is analyzed quantitatively, a permanent continuous circulation of energy can be found, in which the space (often called "vacuum") takes energy from the field and supplies field sources from this energy, which the field sources need in order to continue their emission of field-energy. By the way, it should be mentioned that the Standard model of Cosmology accepts the existence of vacuum-energy, even though Cosmology does not know the origin of this energy.

The central important point of the present work is to show how some energy can be extracted from this energy circulation, and how this energy extracted from the vacuum can be converted into classical mechanical energy. Therefore a special rotor has been developed, so that the theoretical concept was proven and verified experimentally.

The fundamental basics for the practical assembly of a rotor converting vacuum-energy can be found within Electrodynamics:

- Classical Electrodynamics, with the additional information of the finite speed of propagation of the static fields, explains the technical working principle of the vacuum-energy rotor.
- Quantum Electrodynamics, with the additional postulate, that the waves of the zero point oscillations propagate with the same speed as all electromagnetic waves, explains the energy-density of the vacuum, as far as it is concerned with electric and magnetic fields. Concrete values for the energy-density of the electromagnetic zero point oscillations have been calculated in the present work.

The principle has been successfully verified with a measurement of the machine power converted from vacuum-energy !

The practical benefit for the power supply industry free from environmental pollution is obvious: If the principle can be applied on industrial scale, it would not be necessary in future to combust matter in order to supply mankind with energy.

7. References

7.1. External Literature

[Abr 92] The Laser interferometer gravitational wave observatory
Alex Abramovici et.al., 1992. Science 256, S. 325-333

[Ace 02] Status of the VIRGO, F. Acernese et.al., 2002.
Classical and Quantum Gravity 19, S. 1421

[And 02] Current Status of TAMA, M. Ando and the TAMA collaboration, 2002
Classical and Quantum Gravity 19, S. 1409

[Ans 08] Finite Element Program ANSYS, John Swanson (1970-2008)
ANSYS, Inc. Software Products, http://www.ansys.com

[Bar 99] LIGO and the Detection of Gravitational Waves
B. C. Barish und R. Weiss, Phys. Today 52 (Oct 1999), No.10, S. 44-50

[Bec 05] Could dark energy be measured in the lab?
Christian Beck and Michael C. Mackey, Phys. Lett. B, V.605, (2005), p.295.
DOI:10.1016/j.physletb.2004.11.060

[Bec 06] Laboratory tests on dark Energy
Christian Beck, Journal of Physics: Conference Series 31 (2006) 123-130
DOI: 10.1088/1742-6596/31/1/021

[Bec 73] Theorie der Elektrizität, Richard Becker und Fritz Sauter
Teubner Verlag. 1973. ISBN 3-519-23006-2

[Ber 71] Bergmann Schaefer, Lehrbuch der Experimentalphysik, Band 2
Heinrich Gobrecht et.al., Walter de Gruyter Verlag. 1971.
ISBN 3-11-002090-0

[Ber 05] Bergmann-Schäfer Lehrbuch der Experimentalphysik, Band 6, Festkörper
Rainer Kassing et. al., Walter de Gruyter Verlag. 2005.
ISBN 3-11-017485-5

[Bes 07] Structure of the photon and magnetic field induced birefringence and dichroism
J. A. Beswick, C. Rizzo. 2007. arXiv:quant-ph/0702128v1, 13. Feb. 2007

[Bia 70] Nonlinear Effects in Quantum Electrodynamics. Photon Propagation and Photon Splitting in an external Field, Z.Bialynicka-Birula und I.Bialynicki-Birula
Phys. Rev. D. 1970, Vol.2, No.10, page 2341

[Bla 91] Analytic results for the effective action
Steven Blau, Matt Visser and Andreas Wipf
Int. Journ. Mod. Phys. A 6 5408-5434 (1992)

[Boe 07] Exploring the QED vacuum with laser interferometers
Daniël Boer and Jan-Willem van Holten, arXiv:hep-ph/0204207v1, verschiedene Versionen von 17. April 2002 bis 1. Feb. 2008

[Boy 66..08] Timothy H. Boyer has a huge list of publications (Am. J. Phys., Il Nuovo Cimento, Ann. Phys., Phys. Rev., J. Math. Phys., Found. Phys. and many others), beginning in 1966 and going until 2008, it can be found at
http://www.sci.ccny.cuny.edu/physics/faculty/boyer.htm
For many well-known phenomena in quantum theory there are alternative derivations in the framework of stochastic electrodynamics, completely without the use of any quantum theory.

[Boy 80] A Brief Survey of Stochastic Electrodynamics, by Timothy Boyer
Foundation of Radiation Theory and Quantum Electrodynamics
Editor: A. O. Barut, Plenum, New York 1980.

[Boy 85] The Classical Vacuum by Timothy Boyer
Scientific American 253, No.2, August 1985, p.70 -78

[Bre 02] Measurement of the Casimir Force Force between Parallel Metallic Surfaces
G. Bressi, G. Carugno, R. Onofrio, G. Ruoso
Phys. Rev. Lett. 2002. Vol. 88, no.4, S. 4549, S. 041804-1-4

[Bro 28] A Method of and an Apparatus or Machine for Producing Force or Motion
Thomas Townsend Brown, U.S. Patent No. 300,311, 15. Nov. 1928

[Bro 65] Electrokinetic Apparatus
Thomas Townsend Brown, U.S. Patent No. 3,187,206, 1. Juni 1965

[Bru 06] No energy to be extracted from the vacuum
Gerhard W. Bruhn. 2006. *Phys. Scr.* **74** 535-536

[Cal 84..09] Calphysics Institute (Director Bernard Haisch)
There is an extensive list of publications (Ann. Phys., Phys. Rev., Found. Phys. and many others), beginning in 1984 and going until today. It can be found at:
http://www.calphysics.org/index.html

[Cas 48] On the attraction between two perfectly conducting plates.
H. B. G. Casimir (1948), Proceedings of the Section of Sciences
Koninklijke Nederlandse Akademie van Wetenschappen, S.795
as well as H. B. G. Casimir and D. Polder, Phys. Rev. 73 (1948) S. 360

[Cel 07] The Accelerated Expansion of the Universe Challenged by an Effect of
the Inhomogeneities. A Review
Marie-Noëlle Célérier, arXiv:astro-ph/0702416 v2, 7. Jun. 2007

[Che 06] Q & A Experiment to search for vacuum dichroism, pseudoscalar-
photon interaction and millicharged fermions
Sheng-Jui Chen, Hsien-Hao Mei, Wei-Tou Ni
arXiv:/ hep-ex/0611050, 28. Nov. 2006

[Chu 99] Instantaneous Action at a Distance in Modern Physics: Pro and Contra
Andrew E. Chubykalo, Viv Pope, Roman Smirnov-Rueda
Nova Science Publishers. 1999. ISBN-13: 978-1-56072-698-9

[Cod 00] CODATA Recommended Values of the Fundamental Physical
Constants: 1998
Review of Modern Physics 72 (2) 351 (April 2000)
The contents of CODATA are updated continuously at:
http://physics.nist.gov/cuu/Constants/

[Col 93] Extracting energy and heat from the vacuum
Daniel C. Cole and Harold E. Puthoff
Phys. Rev. B, vol.48, no.2, August 1993, p.1562

[Col 96] Foundations of Physics, Vol.26, No.11, 1996, pages 1559-1562
Daniel C. Cole and Alfonso Rueda
Springer Netherlands, ISSN 0015-9018 (Print) 1572-9516 (Online)
DOI: 10.1007/BF02272370

[Cou 99] Bosonic Casimir Effect in External Magnetic Field
Marcus Venicius Cougo-Pinto
Journal of Physics. A: Math. and Gen., Vol. 32, No.24, (1999),
p. 4457-4462
DOI: 10.1088/0305-4470/32/24/310

[Dem 06] Experimentalphysik: Elektrizitat und Optik, Wolfgang Demtröder
Springer Verlag. 2006. ISBN 3-540-33794-6

[Der 56] Direct measurement of molecular attraction between solids separated
by a narrow gap , Boris V. Derjaguin, I. I. Abrikosowa, Jewgeni M.
Lifschitz
Q. Rev. Chem. Soc. 10, 295–329 (1956)

[Dob 06] Physik für Ingenieure, Paul Dobrinski, Gunter Krakau und Anselm Vogel
Teubner Verlag (11.Auflage). 2006. ISBN 978-3-8351-0020-6

[Dow 78] Quantum field theory of Clifford-Klein space-times. The effective Lagrangian and vacuum stress-energy tensor, J. S. Dowker und R. Banach
J. Phys. A: Math. Gen., vol.11, no.11 (1978), S. 2255

[Dub 90] Dubbel - Taschenbuch für den Maschinenbau , 17.Auflage
W. Beitz, K.-H. Küttner et. al., Springer-Verlag. 1990.
ISBN 3-540-52381-2

[Ede 00] Template-stripped gold surfaces with 0.4-nm rms roughness suitable for force
measurements: Application to the Casimir force in the
20-100 nm- range
Thomas Ederth, Phys. Rev. 2000. vol.62, 062104

[Efs 02] The Monthly Notices of the Royal Astronomical Society
George Efstathiou, Volume 330, No. 2, 21. Feb. 2002

[Eng 05] Instantaneous Interaction between Charged Particles
Wolfgang Engelhardt, arXiv:physics/0511172v1, 19.Nov.2005

[Eul 35] Über die Streuung von Licht an Licht nach der Dirac'schen Theorie
H.Euler and B.Kockel. 1935. Naturwissenschaften 23 (1935) 246

[Fey 49a] The Theory of Positrons, Richard P. Feynman
Phys. Rev. 76, No.6 (1949), p.749-759

[Fey 49b] Space-Time Approach to Quantum Electrodynamics, Richard P. Feynman
Phys. Rev. 76, No.6 (1949), p.769-789

[Fey 85] QED. Die seltsame Theorie des Lichts und der Materie, Richard P. Feynman
Übersetzter Nachdruck von 1985, Serie Piper, ISBN: 3-492-03103-X

[Fey 97] Quantenelektrodynamik: Eine Vorlesungsmitschrift, Richard P. Feynman
German translation at Oldenbourg Verlag. 1997. ISBN 3-486-24337-3

[Fey 01] Feynman Vorlesungen über Physik, Band II: Elektromagnetismus und Struktur der Materie, Richard P. Feynman, Robert B. Leighton and Matthew Sands
Oldenbourg Verlag, 3.Auflage. 2001. ISBN 3-486-25589-4

[For 84] Extracting electrical energy from the vacuum by cohesion of charged foliated conductors, by Robert L. Forward, Phys. Rev. B 30, 1700 - 1702 (1984)
DOI: 10.1103/PhysRevB.30.1700

[For 96] An Introductory Tutorial on the Quantum Mechanical Zero Temperature Electromagnetic Fluctuations of the Vacuum." Mass Modification Experiment Definition Study. Philips Laboratory Report #PL-TR 96-3004, (1996), page3

[Ger 95] Gerthsen Physik, H. Vogel
Springer Verlag. 1995. ISBN 3-540-59278-4

[Gia 00] Field correlators in QCD. Theory and Applications
A. Di Giacomo, H. G. Dosch, V. I. Shevchenko, Yu. A. Simonov
arXiv:/hep-ph/0007223 20. Juli 2000

[Gia 06] Physik, Douglas C. Giancoli
Pearson Studium. 2006. ISBN-13: 978-3-8273-7157-7

[Giu 00] Das Rätsel der kosmologischen Vakuumenergiedichte und die beschleunigte Expansion des Universums, Domenico Giulini und Norbert Straumann
arXiv:astro-ph/0009368 v1, 22. Sept. 2000

[Goe 96] Einführung in die Spezielle und Allgemeine Relativitätstheorie, Hubert Goenner
Spektrum Akademischer Verlag, 1996. ISBN 3-86025-333-6

[Gpb 07] Gravity- Probe- B Experiment, Francis Everitt et. al.
to be found in 2008 at http://einstein.stanford.edu/index.html

[Gre 08] Klassische Elektrodynamik, Theoretische Physik (Band 3), Walter Greiner et. al.
Verlag Harri Deutsch. 2008, ISBN 978-3-8171-1818-2

[Hai 08] Quantum vacuum energy extraction, 27. Mai 2008
Bernhard Haisch and Garret Moddel, US Patent #7,379,286

[Hat 81] Proceedings of the First International Symposium on Nonconventional Energy
George D. Hathaway (1981)

[Hec 05] Optik, Eugene Hecht, Oldenbourg-Verlag. 2005. ISBN3-486-27359-0

[Hec 09] Hans Coler: Magnetstromapparat und Stromerzeuger
Explanation in popular scientific language by Andreas Hecht
found in 2009 at: http://www.borderlands.de/energy.coler.php3#2

[Hei 36] Folgerungen aus der Diracschen Theorie des Positrons
W. Heisenberg and H. Euler, 22.Dez.1935,
Zeitschrift für Physik (1936), S.714

[Hei 97] Skript zur Vorlesung „Maschinen- und Feingerätebau für Elektroingenieure"
Prof. J. Heinzl, Technische Universität München

[Hee 00] Nikola Tesla, Collected German and American Patents
von Ulrich Heerd in einem Buch gesammelt im Jahre 2000
Michaels Verlag, MVV Peiting, ISBN 3-89539-246-4

[Hil 96] Elementare Teilchenphysik, Helmut Hilscher
Vieweg Verlag, ISBN 3-528-06670-9

[Hoo 72] Regularization and renormalization of gauge fields
G. 't Hooft and M. Veltman, Nuclear Physics B44 (1972) S.189

[Hur 40] The Invention of Hans Coler, Relating To An Alleged New Source Of Power.
R. Hurst, B.I.O.S. Final Report No. 1043, B.I.O.S.Trip No. 2394
B.I.O.S. Target Number: C31/4799
British Intelligence Objectives Sub-Committee

[Ilm 08] Manufacturer information to the vacuum oil „Labovac 12S"
to be found at
http://www.ilmvac.de/content/products/g11090.html (2008)

[Jac 81] Klassische Elektrodynamik, John David Jackson
Walter de Gruyter Verlag. 1981. ISBN 3-11-007415-X

[Ker 03] Vakuumtechnik in der industriellen Praxis
Jobst H. Kerspe, et. al. 2003, expert-Verlag, ISBN: 3-816-92196-5

[Kho 07a] Experimental test on the applicability of the standard retardation condition to bound magnetic fields, A. L. Kholmetskii, O. V. Missevitch, R. Smirnov-Rueda, R. Ivanov, A. E. Chuvbykalo,
J. Appl. Phys. 101, 023532 (2007)

[Kho 07b] Measurement of propagation velocity of bound electromagnetic fields in near zone, A. L. Kholmetskii, O. V. Missevitch, R. Smirnov-Rueda, R. Ivanov, A. E. Chuvbykalo, J. Appl. Phys. 102, 013529 (2007)

[Kho 08] Note on Faraday's law and Maxwell's equations
Alexander L. Kholmetskii, Oleg V. Missevitch, Tolga Yarman
Eur. J. Phys. 29 (2008) N5–N10 doi:10.1088/0143-0807/29/3/N01

[Kle 08] Photoproduction in semiconductors by onset of magnetic field
Hagen Kleinert and She-Sheng Xue, EPL, 81 (März 2008) 57001

[Kli 03] Elektromagnetischen Feldtheorie, Harald Klingbeil
Verlag Vieweg + Teubner. 2003. ISBN 3-519-00431-3

[Kne 62] Ferromagnetismus, E. Kneller
Springer Verlag. 1962. Library of Congress Catalog
Card Number 62-17383

[Koc 80] Quantum-Noise Theory for the Resistively Shunted Josephson Junction
Roger H. Koch, D. J. Van Harlingen and John Clarke
Phys. Rev. Lett. 45, 2132 - 2135 (1980),
DOI: 10.1103/PhysRevLett.45.2132

[Koc 82] Measurements of quantum noise in resistively shunted Josephson junctions
Roger H. Koch, D. J. Van Harlingen and John Clarke
Phys. Rev. B 26, 74 - 87 (1982), DOI: 10.1103/PhysRevB.26.74

[Köp 97] Einführung in die Quanten-Elektrodynamik, Gabriele Köpp and Frank Krüger
Teubner Verlag, Stuttgart. 1997. ISBN 3-519-03235-X

[Kuh 95] Quantenfeldtheorie, Wilfried Kuhn and Janez Strnad
Vieweg Verlag, Wiesbaden , 1995, ISBN 3-528-07275-X

[Lam 97] Demonstration of the Casimir force in the 0.6 to 6 μm range
Steve K. Lamoreaux, Phys. Rev. Lett., Vol.78, Issue 1, Jan-6-1997, p.5-8
DOI: 10.1103/PhysRevLett.78.5

[Lam 05] Das Vakuum kommt zu Kräften: Der Casimir-Effekt
Astrid Lambrecht, Physik in unserer Zeit, Volume 36 Issue 2,
Pages 85 - 91

[Lam 07] The first axion ?
Steve Lamoreaux, Nature, Vol.441, Seiten 31-32 (2006)

[Lan 97] Lehrbuch der theoretischen Physik, Lew D. Landau and Jewgeni M. Lifschitz Verlag Harri Deutsch (Band 2, Feldtheorie). 1997. ISBN 978-3-8171-1336-1

[Lif 56] The theory of molecular attractive forces between solids.
Jewgeni M. Lifschitz, Sov. Phys. JETP 2, 73–83 (1956)

[Lig 03] Detector Description and Performance for the First Coincidence Observations between LIGO and GEO
The LIGO Scientific Collaboration, 2003, arXiv:gr-qc/0308043 v3, 17. Sept. 2003

[Lin 97] Grundkurs Theoretische Physik, Albrecht Lindner
Teubner Verlag, Stuttgart. 1997. ISBN 3-519-13095-5

[Liu 93] Wormhole created from vacuum fluctuation
Liao Liu, Phys. Rev. D 48 (1993), p.5463 - R5464
DOI: 10.1103/PhysRevD.48.R5463

[Loh 05] Hochenergiephysik, Erich Lohrmann
Teubner Verlag. 2005. ISBN 3-519-43043-6

[Man 93] Quantenfeldtheorie, Franz Mandl and Graham Shaw
Aula-Verlag, Wiesbaden. 1993. ISBN 3-89104-532-8

[Mel 04] Charge Storage in Electret Polymers: Mechanism, Characterization and Applications, Axel Mellinger
Habilitatsionsschrift at the University of Potsdam, 6. Dez. 2004

[Mie 84] Kompendium Hypertechnik. Tachyonenenergie, Hyperenergie, Antigravitation.
Sven Mielordt, Berlin, 1984
Reprint of the 4. edition, raum&zeit Verlag, ISBN 3-89005-005-0

[Moh 98] Precision Measurement of the Casimir Force from 0.1 to 0.9 µm
U. Mohideen and A. Roy, Phys. Rev. Lett. 1998. Vol. 81, no.21, S. 4549

[Mun 09] Measured long-range repulsive Casimir-Lifshitz forces
J. N. Munday, Federico Capasso, V. Adrian Parsegian
Nature 457, 170-173 (8 January 2009) | doi:10.1038/nature07610

[Nie 83] Konversion von Schwerkraft-Feld-Energie. Revolution in Technik, Medizin, Gesellschaft. By Hans A. Nieper
MIT-Verlag, Oldenburg, 1983, 4. erw. Auflage, ISBN 3-925188-00-2

[Ole 99] Faster-than-light transfer of a signal in electrodynamics
Valentine P. Oleinik
Published in web:
www.chronos.msu.ru/EREPORTS/oleinik_faster.pdf
And published at: Instantaneous action-at-a-distance in modern physics
Nova Science Publishers, Inc., New York, 1999, p. 261-281.

[Pas 99] Programmiersprache PASCAL, Compiler from 1999

[Pau 00] Relativitätstheorie, Wolfgang Pauli
Reprint by Springer-Verlag, 2000. ISBN 3-540-67312-1

[Pen 96] The Quantum Dice: An Introduction to Stochastic Electrodynamics
Luis de la Pena, Ana Maria Cetto, Kluwer Academic Publishers, 1996
Dordrecht, The Netherlands, ISBN 0-7923-3818-9

[Per 98] Measurements of Ω and Λ from 42 high-redshift Supernovae
S. Perlmutter, et. al., arXiv:astro-ph/9812133, 8. Dez. 1998

[Pin 99] Engine cycle of an optically controlled vacuum energy transducer
Fabrizio Pinto, Phys. Rev.B, 60, 21, 1999, p. 14740-14755
DOI: 10.1103/PhysRevB.60.14740

[Pin 03] Fabrizio Pinto, US patents #6,650,527 and #6,665,167 #6,477,028
Fabrizio Pinto also founded the InterStellar Technologies Corporation for the investigation of the utilization of the energy of the vacuum:
http://www.interstellartechcorp.com/companyFabrizio.html

[Pop 00] Vacuum polarization of a scalar field in wormhole spacetime
Arkadii A. Popov und Sergey V. Sushkov
arXiv:gr-qc/0009028v1 10. Sept. 2000

[Pop 05] Long throat of a wormhole created from vacuum fluctuation
Arkadii A. Popov, Class. Quantum Grav. 22 (2005) p.5223-5230
DOI: 10.1088/0264-9381/22/24/002

[Put 08] Linearized Turbulent Fluid Flow as an Analog Model for Linearized General Relativity (Gravitoelectromagnetism)
H. E. Puthoff, arxiv.org/abs/0808.3404, 1. Sept. 2008

[Red 08] Comment on 'Note on Faraday's law and Maxwell's equations'
Dragan V Redzic, Eur. J. Phys. 29 (2008) L33–L35

[Rie 98] Observational Evidence from Supernovae for an Accelerating Universe and a Cosmological Constant, Adam G. Riess et. al.
arXiv:astro-ph/9805201, 15. Mai. 1998

[Rik 00] Magnetoelectric birefringences of the quantum vacuum
G. L. J. A. Rikken and C. Rizzo, 2000, Phys. Rev. A, Vol.63, 012107 (2000)

[Rik 03] Magnetoelectric anisotropy of the quantum vacuum
G. L. J. A. Rikken and C. Rizzo, 2003, Phys. Rev. A, Vol.67, 015801 (2003)

[Riz 07] The BMV project: Biréfringence Magnétique du Vide
Presentation at March-15-2007 by Carlo Rizzo at the meeting „Rencontres de Moriond", to be found at:
http://moriond.in2p3.fr/J07/sched07.html

[Sch 49] Quantum Electrodynamics. III. The Electromagnetic Properties of the Electron – Radiative Corrections to Scattering
Richard P. Feynman, Phys. Rev. 76, No.6 (1949), p.790-817

[Sch 88] Feynman-Graphen und Eichtheorien für Experimentalphysiker,
Peter Schmüser
Springer Verlag. 1988. ISBN 3-540-58486-2

[Sch 02] Gravitation, Ulrich E. Schröder
Verlag Harri Deutsch, 2002. ISBN 3-8171-1679-9

[Sch 07] Stephan Schiller, private communication, 2007
University of Düsseldorf, Germany

[Sha 83] Black Holes, White Dwarfs and Neutron Stars: The Physics of Compact Objects
Stuart L. Shapiro and Saul A. Teukolsky
Wiley Interscience. 1983. ISBN 0-471-87317-0

[She 01] Possible vacuum-energy releasing
She-Sheng Xue, Physics Letters B 508 (2001) 211-215

[She 03] Magnetically induced vacuum decay
She-Sheng Xue, Phys. Rev. D 68 (2003) 013004

[Sim 80] Absolute electron-proton cross sections at low momentum transfer measured with a high pressure gas target system
G.G.Simon, Ch.Schmitt, F.Borkowski, V.H.Walther
Nucl. Phys. A, vol.333, issue 3, (1980), p.381-391

[Sim 04] Kulturgeschichte der Physik (hier speziell Abschnitt 2.5.)
Károly Simonyi, Verlag Harri Deutsch. 2004. ISBN 3-8171-1651-9

[Sol 03] Some remarks on Dirac's hole theory versus quantum field theory
Dan Solomon, Can. J. Phys. 81: 1165-1175 (2003)

[Sol 05] Some differences between Dirac's hole theory and quantum field theory
Dan Solomon, Can. J. Phys. 83: 257-271 (2005)

[Sol 06] Some new results concerning the vacuum in Dirac's hole theory
Dan Solomon, Phys. Scr. 74 (2006) 117-122

[Spa 58] Measurement of attractive forces between flat plates.
Marcus J. Spaarnay, Physica. Band 24, 1958, S. 751.

[Spr 85] Vorrichtung oder Verfahren zur Erzeugung einer variirenden Casimiranalogen Kraft und Freisetzung von nutzbarer Energie
Werner Sprenzel and Sven Mielordt
German Patent, Aktenzeichen DE3541084, announced at 15.11.1985.

[Stö 07] Taschenbuch der Physik, Horst Stöcker
Verlag Harri Deutsch. 2007. ISBN-13 987-3-8171-1720-8

[Sve 00] Precise Calculation of the Casimir Force between Gold Surfaces
V. B. Svetovoy and M. V. Lokhanin
Modern Physics Letters A. 2000. vol.15, nos. 22 & 23, S.1437

[Teg 02] Measuring Spacetime: from Big Bang to Black Holes
Max Tegmark, arXiv:astro-ph/0207199 v1, 10. Juli 2002
Slightly abbreviated version in: Science, 296, 1427-1433 (2002)

[Tes 91] Nikola Tesla, addressing the American Institute of Electrical Engineering, 1891

[Thi 18] Über die Wirkung rotierender ferner Massen in Einsteins Gravitationstheorie
Hans Thirring und Josef Lense, Phys. Zeitschr. 19, Seiten 33-39 Jahrgang 1918

[Tip 03] Moderne Physik, P. A. Tipler und R. A. Llewellyn
Oldenbourg Verlag. 2003. ISBN 3-486-25564-9

[Ton 03] Cosmological Results from High-z Supernovae
John L. Tonry, et. al., arXiv:astro-ph/0305008, 1. Mai 2003

[Tur 07] Prüfungstrainer Physik (textbook for students), Claus W. Turtur
B. G. Teubner Verlage, Wiesbaden. 2007. ISBN 978-3-8351-0137-1

[Umr 97] Grundlagen der Vakuumtechnik
Walter Umrath, Leybold Vakuum GmbH, 1997
to be found at
www-ekp.physik.uni-karlsruhe.de/~bluem/grund_vac.pdf (2008)

[Val 05] Practical Conversion of Zero Point Energy: Feasibility Study of Zero Point Energy Extraction from the Quantum Vacuum for the Performance of Useful Work
by Thomas Valone, Integrity Research Institute, 2005.
ISBN 978-09641070-8-3

[Val 08] Zero Point Energy, The fuel of the future
Thomas Valone, Integrity Research Institute, 2008.
ISBN 978-0-9641070-2-1

[War 09] Externally-shunted high-gap Josephson junctions: Design, fabrication and noise measurements, EPSRC grant EP/D029872/1
P.A. Warburton, grant from 01. April 2006 to 31. March 2009
http://gow.epsrc.ac.uk/ViewGrant.aspx?GrantRef=EP/D029783/1

[Whe 68] Einsteins Vision
Wie steht es heute mit Einsteins Vision, alles als Geometrie aufzufassen?
John Archibald Wheeler. 1968. Springer Verlag

[Wik 08] http://de.wikipedia.org/wiki/Elektret und
http://de.wikipedia.org/wiki/Anorganisch
Allgemeine Information über Elektrete und über anorganische Elektrete.

[Wik 09] Design of the „Stromerzeuger" according to Hans Coler:
http://expliki.org/wiki/Coler-Energie-Konverter
Design of the „ Magnetstromapparat" according to Hans Coler:
http://expliki.org/wiki/Magnetstromapparat

[Wil 02] A report of the status of the GEO 600 gravitational wave detector, 2002
B.Willke et. al., Classical and Quantum Gravity. 19, S.1377

[Zav 06] Experimental Observation of Optical Rotation Generated in Vacuum by a Magnetic Field, E.Zavattini et. al., Phys. Rev. Lett. 96, 110406 (2006)

[Zav 07] New PVLAS results and limits on magnetically induced optical rotation and ellipticity in vacuum, E.Zavattini, et. al., arXiv:0706.3419v2 , 26. Sept. 2007

7.2. Own publications in connection with the present work

[e1] Does cosmological vacuum energy density have an electric reason ?
Claus W. Turtur: http://arXiv.org/abs/astro-ph/0403278 (März 2004)

[e2] A Theoretical Determination of the Electron's Mass
Claus W. Turtur, Galilean Electrodynamics & GED East, Volume 17, Special Issues 2, Fall 2006, S.23-29

[e3] Systematics of the Energy Density of Vacuum Fluctuations and Geometrodynamical Excitones, Claus W. Turtur, Physics Essays, Vol#20, No.2 (Juni 2007)

[e4] About the Electrostatic Field following Coulomb's law with additional Consideration of the finite speed of propagation following the theory of Relativity, Claus W. Turtur, PHILICA.COM, ISSN 1751-3030, Article number 112 (11. December 2007)

[e5] Two Paradoxes of the Existence of electric Charge
Claus W. Turtur, arXiv:physics/0710.3253 v1 (Okt.2007)

[e6] A QED-model for the Energy of the Vacuum and an Explanation of its Conversion into Mechanical Energy, Claus W. Turtur
PHILICA.COM, ISSN 1751-3030, Article number 138, (4. Sept. 2008)

[e7] A Hypothesis for the Speed of Propagation of Light in electric and magnetic fields and the planning of an Experiment for its Verification
Claus W. Turtur: http://arXiv.org/abs/physics/0703721
Version Nr.2, Nov. 2007.

[e8] Conversion of vacuum-energy into mechanical energy: Successful experimental Verification, Claus W. Turtur
PHILICA.COM, ISSN 1751-3030, Article number 124, (2. April 2008)

[e9] Conversion of vacuum-energy into mechanical energy
Claus W. Turtur, The General Science Journal, ISSN 1916-5382 (5. Juni 2008)
Im Internet abrufbar unter http://wbabin.net/physics/turtur.pdf

[e10] Two Paradoxes of the Existence of magnetic Fields, Claus W. Turtur
PHILICA.COM, ISSN 1751-3030, Article number 113, (19. December 2007)

[e11] A Motor driven by Electrostatic Forces, Claus W. Turtur
PHILICA.COM, ISSN 1751-3030, Article number 119, (18. Februar 2008)

[e12] An electrostatic rotor with a mechanical bearing, Claus W. Turtur
PHILICA.COM, ISSN 1751-3030, Observation number 45, (11. April 2008)

[e13] The role of Ionic Wind for the Electrostatic Rotor to convert Vacuum Energy into Mechanical Energy, Claus W. Turtur
PHILICA.COM, ISSN 1751-3030, Observation number 49, (16. Sept. 2008)

[e14] Conversion of Vacuum Energy into Mechanical Energy under Vacuum Conditions
Claus W. Turtur,
PHILICA.COM, ISSN 1751-3030, Article number 141, (3. Dez. 2008)

[e15] A magnetic rotor to convert vacuum-energy into mechanical energy, Claus W. Turtur
PHILICA.COM, ISSN 1751-3030, Article number 130, (21. Mai 2008)

[e16] An easy way to Gravimagnetism
Claus W. Turtur: http://arXiv.org/abs/physics/0406078 (Juni 2004)

[e17] Definite Proof for the Conversion of vacuum-energy into mechanical energy based on the measurement of machine power, Claus Turtur and Wolfram Knapp
PHILICA.COM, ISSN 1751-3030, Article number 155, (2. April 2009)

[e18] Conversion of the Vacuum-energy of electromagnetic zero point oscillations into Classical Mechanical Energy
Claus W. Turtur, The General Science Journal, ISSN 1916-5382 (5. Mai 2009)
In Internet abrufbar unter http://wbabin.net/physics/turtur1e.pdf

7.3. Cooperations and private communication

[Bec 08/09] Hans Peter Beck
Technical University of Clausthal and Energieforschungszentrum Niedersachsen
Some of his co-workers supported my work in the years 2008 and 2009 at my experimental investigation in the laboratory.

[Ihl 08] Private communication, Thank goes to W. Guilherme Kürten Ihlenfeld of the Physikalisch Technische Bundesanstalt Braunschweig, who allowed me, to use the ANSYS-program on his computer.

[Kah 08] Martin Kahmann and W. Guilherme Kürten Ihlenfeld
They have been the first colleagues who suggested me to verify the criterion of "net energy gain", also known as "over-unity" criterion. They made this suggestion on the occasion of a demonstration, which I made in April 2008 to present my electrostatic rotor at the Physikalisch Technische Bundesanstalt (PTB) in Braunschweig. With their suggestion they defined a goal for me, for I worked in the time after this demonstration.

[Kna 08/09] Wolfram Knapp
These experimental investigations in the vacuum have been done in cooperation with the "Institut für Experimentelle Physik" of the Otto-von-Guericke-University at Magdeburg.

[Lam 09] Steven Lamoreaux from Yale-University, U.S.A.
In a thought-exchange per Email, Prof. Lamoreaux suggested me to bring the experiment of the vacuumenergy-rotor down to a temperature as close as possible to the absolute T = 0K, in order to invesatigate, whether the rotor is driven by thermal radiation or by electromagnetic zero-point radiation.

[Lie 08/09] Frank Lienesch, Physikalische Technische Bundesanstalt Braunschweig (PTB). Here I got repeatedly the possibility to support my little transportable cryostate for tests on a magnetic rotor with liquid nitrogen, so that I could conduct experiments with superconducting rotors in magnetic fields.

[Ost 07] G.-P. Ostermeyer and J.-H. Sick from the University of Braunschweig gave me the possibility to use an interferometer for experiments regarding the speed of propagation of light in electrostatic fields. I could insert my setup into their interferometer and do some measurements which I published in [e7].

Zeitfracht Medien GmbH
Ferdinand-Jühlke-Straße 7
99095 Erfurt, Deutschland
produktsicherheit@kolibri360.de